Galloping the Globe

The Geography Unit Study for Young Learners

Kindergarten through 4th grade

by Loree' Pettit & Dari Mullins

Dedication

Galloping the Globe is lovingly dedicated to our children, Jason and Aaron Pettit and Artie, Autumn, and Aspen Mullins. The five of you are the biggest blessings in our lives. We are so thankful that God has given us the privilege of being your moms.

Acknowledgements

Loree'-

I would like to thank God for enabling me to do more than I ever dreamed possible. Three years ago when I was at the lowest point of my life, You promised to bring beauty from ashes. True to your nature, You have done so abundantly, exceedingly more than I ever thought. To my husband and best friend, Ralph, thank you for being there and encouraging me. God definitely knew what He was doing when He put us together. To my heart sister, Dari, you are one of the "abundantly, exceedingly more than I ever thought" gifts from God. Thank you for all of your hard work on this book. But mostly, thank you for your friendship.

Dari-

I would like to thank the Lord for His providential hand in my life. Thank You for directing my paths and leading me to places I never dreamed possible. To my husband, Allen, for loving me in spite of myself and for being the best friend and father anyone could ever have. To my best friend and sister-in-the-Lord, Loree', thank you for always being there. You are the epitome of what God teaches us to be; I appreciate your example. I also want to thank you for taking the idea of publishing this course and running with it. I am sure you can type at least 500 wpm by now. You deserve most of the credit for this book!

We would like to extend a very special thank you to Letz Farmer of Mastery Publications for the advice and time she gave us in the development of *Galloping the Globe*. Shortly after the completion of *Galloping the Globe*, Letz was diagnosed with a rare form of cancer and passed away. Letz was a truly remarkable person that left a lasting impact on those around her. She will not be forgotten.

Published by Geography Matters, Inc.

Edited by Cindy Wiggers
Book layout by Libby Wiggers

Programs utilized during production:
Serif Page Plus 8
Puzzle Power
Art Explosion
ClipTures volume 4

ISBN 1-931397-21-x

Library of Congress Control Number: 2003102121

Printed in the United States of America

Geography Matters, Inc.
800.426.4650
www.geomatters.com

Galloping the Globe
Table of Contents

Objective

The main objective of *Galloping the Globe* is to introduce children to the seven continents and some of the countries within each continent. Along with studying geography, the students will be introduced to various historical figures, missionaries, and animals of the world. This study is geared for the kindergarten to fourth grade level. It can be adapted to older children; however, the majority of the books listed under each section are for the early elementary age group.

How This Program Works

Galloping the Globe covers eight divisions:

❖ **People/History** – This section covers missionaries, historical figures, authors, artists, musicians, scientists, etc.

❖ **General Reference Books** - These books cover general information about a country or topic.

❖ **Literature** – Children's literature originating from the country, set in the country, or about the country. This section includes classic children's literature and some titles that were chosen just because they are fun.

❖ **Science** – The main topic of this section is animals that are native to a country or are commonly found in a country. Other topics include: trees, volcanoes, astronomy, flowers, fruits and vegetables, rain forest, and the coral reef. In South America and Africa, science is listed under the continent rather than the individual countries.

❖ **Vocabulary** – On page 232 you will find a "Geography Dictionary" page. This is a blank page that you may photocopy as many times as needed for the students in your immediate family to make their own dictionaries. Using this template, students will define and draw a picture of the geography vocabulary. You may want to make a dictionary per country or continent. Defining vocabulary words provides a great opportunity to introduce dictionary skills. While organizing the "Geography Dictionary" your students will have natural reasons to work on using alphabetical order.

❖ **Music/Art/Projects** - This section includes a variety of crafts using readily available materials. Also included are recipes from that country, music to enjoy, and other activities pertaining to the country's culture.

❖ **Internet Resources** - Included in this section are web sites with information, pictures, activities, recipes, and other interesting facts.

❖ **Bible** - Most of the scripture included in this section is related to character traits found in the literature books. Some of the verses pertain to items listed in the Science section or other areas related to that country's culture.

This course is designed to be completed in one to three school years. To complete it in one school year, you will not be able to study every country, read every book, do every activity, or study every science topic listed. It will take two to three school years to cover every topic listed, depending upon how long you spend with each topic.

Suggestions:

- Read one to three of the literature books on the first day of each country. This will help make the children curious about that country. With some students you may need to read a literature book a day to keep that curiosity going.

- Choose the activities and books that are right for your family. Do an overview of the continent and study three to five countries per continent. While most countries are divided into the eight activity sections, some countries have fewer sections, as appropriate resources were not as readily available.

- Books listed under each section are only suggestions. Your public library or personal bookshelves may have other wonderful books on the subject. In most sections there are more books suggested than you really need. With the possible exception of literature, choose one to three books per topic.

- Some of the books or websites may make reference to the earth being millions of years old or other evolutionary statements. We suggest you review the books before reading them to your children and use your own discernment. You can skip the evolutionary statements, change them around, or if your child is old enough and you feel is mature enough, you can discuss the statements in light of what Scripture says. We tried not to list books that were, in our opinion, "heavy" with evolution. If any got by us, we apologize.

- A few of the books listed in the Literature section also reference other gods and/or religions. This can be handled in several ways: skip the book entirely, leave out that section of the book, or discuss the topic in light of Scripture. Older children can begin to understand the importance of missionaries and they may add that country to their prayer list. We tried to weed out the books with an abundance of these references but included some because the rest of the information in the books were of significant value. Again, use your discernment to determine what is suitable for your family.

Getting Started

Before you begin be sure to have a children's atlas, children's dictionary, and a globe readily available. In addition you will be using a variety of books. Most of the books used in this program were found in our local library. Check your library for availability. There are a few resources that you may want to buy since they are used throughout the course. Besides the convenience of having them handy, most can also be used for years to come. These foundational books are:

- ◆ *Considering God's Creation* by Sue Mortimer & Betty Smith
- ◆ *Our Father's World* (second grade social studies by Rod & Staff)
- ◆ *Missionary Stories with the Millers* by Mildred A. Martin
- ◆ *Heaven's Heroes* by David Shibley
- ◆ *Baby Animal Stories* – A compilation published by Publications Int'l, Ltd. ISBN 0-7853-2680-4
- ◆ *Special Wonders of the Wild Kingdom* by Buddy & Kay Davis
- ◆ *The Usborne Book of Famous Lives* by Struan Reid & Patricia Fara
- ◆ *Children Just Like Me* by Barnabas and Anabel Kindersley

Building Student Notebooks

Galloping the Globe utilizes the notebook approach on a very basic level. In the notebook approach, the child builds a notebook over the length of the course. This encourages students to be producers rather than consumers.

Obtain a 2" or 3" three-ring binder and dividers for each student. Use dividers to separate the student notebook into the following sections:

- Bible
- Handwriting
- Introduction to Geography
- Asia
- Europe

- Christmas Around the World
- The Poles
- North America
- South America
- Africa

- Australia
- Vocabulary
- Butterflies
- Insects
- Art

Art is for those wonderful creations that have nothing to do with any of the aforementioned topics. As you study a country or subject, put the child's work into the appropriate section. (A three-hole punch comes in handy.)

Notebook Suggestions are given throughout this program, but feel free to use your own ideas as well. Use the reproducible sheets provided in the back of this book for a geography dictionary, biographies, country studies, and animal reports. At the end of the year, the child will have a finished book that they can show relatives and friends, reinforcing and reviewing all that they have learned throughout the year. The children can refer to their notebook, solidifying what they have learned in addition to building confidence in themselves for the progress they have made.

Other Resource Recommendations

For a complete year-long course in U.S. and World Geography for your older students we recommend *Trail Guide to World Geography* and *Trail Guide to U.S. Geography*. These books use the notebook approach to learning and include a myriad of hands-on projects. For more outline maps beyond what is provided in this program, *Uncle Josh's Outline Map Book* or CD-ROM is a handy resource for studying geography, history, and more.

A geographical terms chart is a valuable reference tool to use while making the "Geography Dictionary". It is a composite drawing of dozens of physical features of the earth with the geography terms labeled right on the picture. The reverse side of the chart includes basic definitions of over 150 terms. To obtain a laminated terms chart see Resources on page 240.

Answer Key

Answers to notebook questions and crosswords are provided in the Answer Key located in the Appendix. All efforts have been made to assure accuracy. Please feel free to contact the publisher if any corrections are needed.

Giddyap, Let's Go!

If you have obtained the necessary resources, you're ready to roll. Follow our mascot, Gulliver, as he gallops the globe through the pages of this book. May this curriculum serve to make your study of geography a pleasure and spur your students to develop a life-long interest in the people and culture of God's wide world.

Introduction to Geography

Using a Map

Begin this unit by covering basic map and globe skills.

- ❑ *Maps and Globes* by Jack Knowlton
- ❑ *Maps and Mapping* by Barbara Taylor
- ❑ *How to Use a Map* by Evan-Moor Publishers
- ❑ *Globes* (A New True Book) by Paul P. Sipiera
- ❑ Pages 41-42 *Our Father's World*
- ❑ Pages 46-48 *Our Father's World*
- ❑ Pages 96-98 *Our Father's World*

Children should know north, south, east, and west and the difference between land and water. *Katy and the Big Snow* by Virginia Lee Burton is an excellent children's book to introduce them to direction.

Notebook Suggestions:

- ❑ *How to Use a Map* by Evan-Moor Publishers
- ❑ Make a map of your house, neighborhood, or a park.

Basic Geography

Geography – A description of the earth or globe.

Christian geography is the view that the earth's origins, ends, and purposes are of Christ and for His glory.

- ❑ Continents
  - *Continents* (A New True Book) * by Dennis B. Fradin (Note: evolutionary content)
  - Pages 37-40 *Our Father's World*

- ❑ Oceans
  - *The Pacific Ocean* (A New True Book) * by Susan Heinrichs
  - *The Atlantic Ocean* (A New True Book) * by Susan Heinrichs
  - *The Indian Ocean* (A New True Book) * by Susan Heinrichs
  - Pages 43-45 *Our Father's World*

*These books can be a bit lengthy to hold a young elementary student's attention. We recommend just covering the highlights.

People/History

❑ Mercator
  • Page 7 - *The Usborne Book of Famous Lives*

❑ Explorers
  • Page 99 - *The Usborne Book of Famous Lives*
  • *Explorers* (A New True Book)
  • *Explorers and Adventurers* – Child's First Library of Learning

❖ Christopher Columbus
  • Pages 118-119 – *Usborne Book of Famous Lives*
  • *Christopher Columbus* by Stephen Krensky – Step into Reading
  • *Christopher Columbus* by Tanya Larkin
  • *Christopher Columbus* – Young Christian Library
  • *Follow the Dream* by Peter Sis
  • *Columbus* by D'Aulaire
  • Christopher Columbus – Animated Hero Classics Video

❖ Francis Drake
  • Page 135 – *Usborne Book of Famous Lives*
  • *Sir Francis Drake* by Tanya Larkin
  • *Francis Drake* by David Goodnough

❖ Leif Ericson
  • Page 117 – *Usborne Book of Famous Lives*
  • *Leif the Lucky* by D'Aulaire

❖ Cabeza de Vaca
  • *Cabeza de Vaca: New World Explorer* by Keith Brandt

❖ Vasco da Gama
  • *Vasco da Gama* by Tanya Larkin

❖ Hernando de Soto
  • *Hernando de Soto* by Tanya Larkin

❖ Have the child write a creative story of himself/herself as an explorer.

❖ Have the child draw a picture of an explorer or of himself/herself as an explorer.

General Reference Books

- ❑ Pages 100-113 *Our Father's World*

Literature

- ❑ *How To Make An Apple Pie and See the World* by Marjorie Priceman
- ❑ *Miss Rumphius* by Barbara Cooney
- ❑ *Katy & the Big Snow* by Virginia Lee Burton
- ❑ *Amelia's Fantastic Flight* by Rose Bursik
- ❑ *Letters from Felix* by Annette Langen (There is an activity book that goes with this.)

Science

God has a purpose for the Earth. The Earth is part of the universe and one of nine known planets that circle the sun.

- ❑ Lesson 2 – *Considering God's Creation*

- ❑ Insects
  - *God made the Firefly* – God is Good Series by Rod & Staff
  - *Creepy Crawlies* by Wendy Madgwick
  - *Bugs for Lunch* by Margery Facklam
  - *Monster Bugs* by Lucille Recht Penner
  - *Insects* – Golden Book – ISBN: 0-307-20400-6
  - *Little Honeybee* - Baby Animal Stories ISBN: 0-7853-2680-4
  - *Honey Bees and Hives* by Lola M. Schaefer
  - *Bumble Bees* by Cheryl Coughlan
  - *Fireflies* by Cheryl Coughlan
  - *Mosquitoes* by Cheryl Coughlan
  - *Ladybugs* by Cheryl Coughlan
  - *Pets in a Jar* by Seymour Simon
  - *The Very Quiet Cricket* by Eric Carle
  - *The Very Lonely Firefly* by Eric Carle
  - *The Icky Bug Counting Book* by Jerry Pallotta
  - *Bugs! Bugs! Bugs!* by Bob Barner
  - Lesson 13 – *Considering God's Creation*
  - Play the "Bug Game" by Ampersand Press
  - Make an ant colony.

Notebook Suggestions:

1. List the three main body parts of an insect. Draw an insect and label the parts.

2. Describe the function of the hard covering on the outside of an insect's body.

3. Explain how insects use their jaws.

4. Explain what makes ants social insects.

5. Where do ants do NOT live?

6. What is a group of ants called?

7. Draw a picture of an ant colony.

8. Describe how male crickets chirp.

9. Explain how crickets hear.

10. List insects that live around your home. Include such information as description, habitat, life span, food.

11. List insects that are helpful to gardens. Explain how or why.

12. List insects that are harmful to gardens. Explain how or why.

13. Draw or color pictures of at least two insects.

❑ Butterflies

- *Butterflies* by Emily Neye
- *Butterflies and Moths* Usborne First Nature
- *Amazing World of Butterflies and Moths* by Louis Sabin (Except for evolutionary statements on page 5, this a wonderful book. If you choose to read it, just skip page 5.)
- *Monarch Butterflies* by Helen Frost
- *Butterfly Colors* by Helen Frost

Notebook Suggestions:

1. List four stages of a butterfly's life.

2. List the three main body parts of a butterfly.

3. What is the antennae used for?

4. With what are the wings covered?

5. What do caterpillars eat?

6. What do butterflies eat?

7. Find and identify butterflies around your home. Include a drawing or picture in your notebook.

8. Draw or color pictures of 1 or 2 butterflies.

9. Catch butterflies or make a butterfly habitat. *Butterfly Jungle* by Uncle Milton provides the habitat and a coupon that you send off for live caterpillars. Include a picture or report of this project in your notebook.

Vocabulary

map	compass	insect
cartography	north	bug
globe	south	head
explorer	east	thorax
	west	abdomen

Music/Art/Projects

1. Make a paper mache earth.

2. If you read *How to Make an Apple Pie and See the World*, bake an apple pie.

3. If you studied insects, make ice cream ants.

vanilla ice cream	Magic Shell chocolate coating	pretzel sticks
M&M's	shoe string licorice	

- On a plate, place three small scoops of ice cream in a line.
- Place two M&Ms on the "head" for the eyes. Cut an M&M in half. Use the two halves as the mouth. Use the licorice as the antennae.
- Poke three pretzel sticks per side in the middle section or "thorax".
- Cover the whole "ant" with Magic Shell as the "exoskeleton".

Internet Sources

- ❑ http://www.eduplace.com/ss/ssmaps/index.html
- ❑ http://www.yourchildlearns.com/megamaps (not compatable with Macintosh)

Bible

❑ *How to Make an Apple Pie and See the World*
- Proverbs 25:11
- Psalm 113:4
- Genesis 1:1

❑ *Katy and the Big Snow*
- I Peter 5:6
- Genesis 39:19-41:13

❑ *Miss Rumphius*
- Matthew 13:1-8
- Mark 4:3-8
- Luke 8:4-8
- Matthew 13:31-32
- Mark 4:30-32
- Luke 13:18-19
- Genesis 8:22
- 2 Corinthians 9:10
- Galatians 6:7
- Mark 4:26-29
- John 1:1,3,14

❑ God is responsible for the contour of the earth.
- Genesis 1:1-8
- Isaiah 48:13
- Isaiah 40:22
- Nehemiah 9:6
- Job 28:9-11
- Job 26:7-12

❑ God controls His creation.
- Psalm 107:23-31
- Job 36:26-28
- Genesis 1:9-25
- Psalm 24:1
- Psalm 113:4

❑ God originated nations and languages.
- Isaiah 45:18
- Genesis 11:1-9
- Job 12:23
- Acts 17:24-28
- Genesis 10:5

❑ God is concerned with all people.
- Acts 10:34
- Romans 10:12, 13
- John 4
- Matthew 28:19
- Acts 1:8
- Acts 2:5-11

Continents

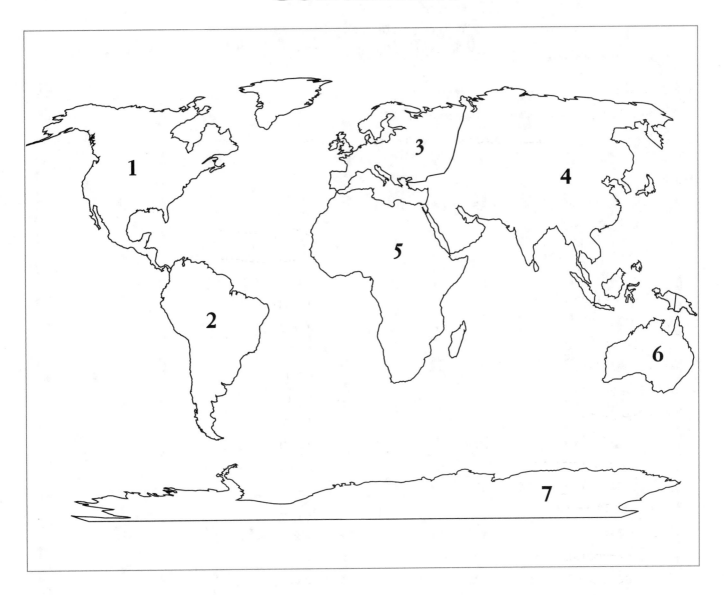

1._____

2._____

3._____

4._____

5._____

6._____

7._____

Word List

Africa
Asia
Oceania (Australia)
Europe
North America
South America
Antarctica

Word Search

```
v  u  g  j  i  c  i  g  l  o  b  e
e  a  s  t  y  o  b  e  t  y  l  k
i  u  g  u  l  l  s  x  i  i  f  t
m  c  b  i  f  u  m  p  i  n  t  a
a  t  c  n  l  m  l  l  r  a  s  p
p  t  v  o  f  b  e  o  m  e  j  c
p  b  d  r  n  u  a  r  f  n  i  o
w  e  s  t  o  s  s  e  h  z  d  m
s  y  z  h  f  n  f  r  q  s  x  p
s  o  u  t  h  d  c  y  u  l  j  a
n  d  q  k  o  c  e  a  n  n  d  s
n  i  n  a  s  i  q  m  r  b  q  s
```

Columbus
compass
east
explorer
globe
map

Nina
north
ocean
Pinta
south
west

14

Word Search

```
q p c o n t i n e n t g
q a s i a z z a x f s x
r f h o r x j n o r t h
s r z k x k b t r f s z
r i b m k k p a z r s v
z c x w v z x r h x o e
c a j l e u k c j d u e
n o x p u q u t a l t w
a u s t r a l i a s h w
s a t y o m b c a m s e
f l u y p x j a c a c e
h h a m e r i c a j l h
```

Africa
America
Antarctica
Asia
Australia

Continent
Europe
North
South

MAZE CRAZE

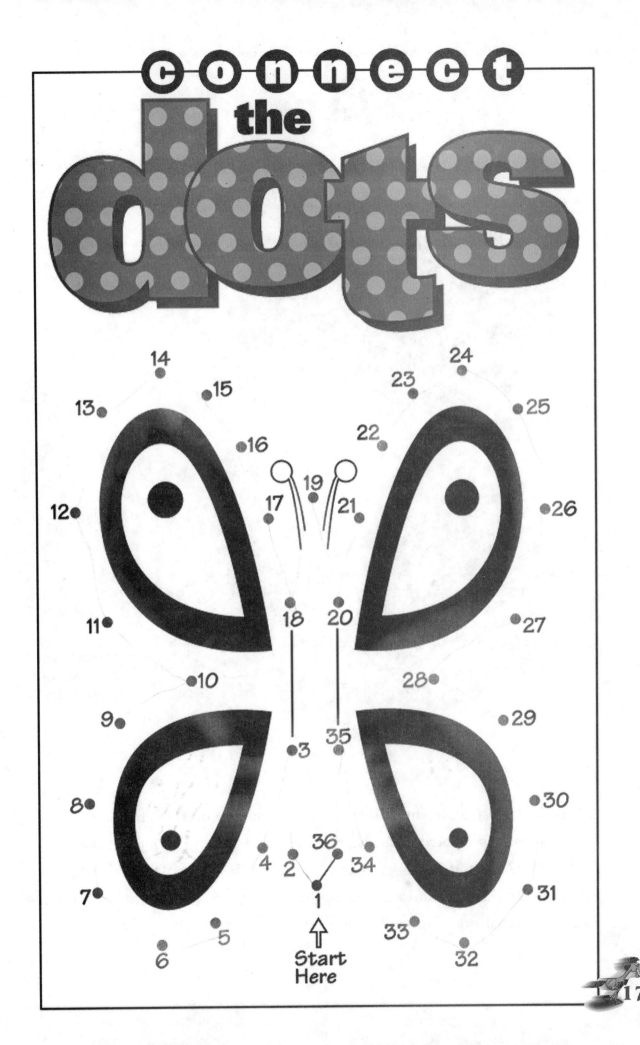

Asia

Asia is the largest continent in the world and home to over 3.7 billion people. Its two highest populated countries are China and India. The highest point in Asia is 29,028 feet at Mt. Everest in China (Nepal) - Tibet. The lowest elevation is at the Dead Sea in Israel, Jordan where it is 1,339 feet below sea level. Asia is separated from Europe by the Ural Mountains.

Map of Asia

Color each country you study.

Russia (Europe)

Ural Mtns.

Russia

Turkey

Israel

Iraq

Kazakhstan

Mongolia

North Korea

Japan

Iran

South Korea

Saudi Arabia

China

Pacific Ocean

Pakistan

Nepal

India

Thailand

Indian Ocean

Indonesia

© 2003 Geography Matters

ad maiorem Dei gloriam!

19

China

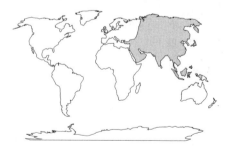

20

Population_____

Capital City_____

Religion_____

Type of Government_____

Currency_____

Language_____

What are the people called?_____

China

Over two-thirds of Chinese workers are farmers. China's main products are rice, wheat, cotton, tobacco, and silk. Rice is a main part of the Chinese diet. Millions of tons of rice must be produced each year to feed the vast population. China is the largest producer of rice, tobacco, and cotton in the world.

Riding bicycles is a popular means of transportation in China. There are over 130 million bicycles in China and every city has bicycle parks with attendants to look over them.

Some of the common items we use today were invented in China. They include kites, paper money, wheelbarrows, porcelain, and folding umbrellas. A variety of foods discovered in China and introduced to the West are apricots, peaches, oranges, grapefruits, and tea.

Stretching about 4,000 miles, the Great Wall of China was built to defend the country from northern attacks. The Great Wall took over 25 years to build. It is the only man-made structure that can be seen from a spacecraft 200 miles above the earth.

People/History

- ❑ Hudson Taylor
  - *Hudson Taylor* – Young Christian Library
  - Too Busy Fishing – chapter 8, *Missionary Stories with the Millers*

- ❑ Eric Liddell
  - *Eric Liddell* – Young Christian Library
  - Flying Scot – chapter 8, *Heaven's Heroes*

- ❑ Marco Polo
  - Pages 112-113 – *Usborne Book of Famous Lives*
  - *Marco Polo* by Gian Paolo Ceserani

- ❑ Great Wall of China
  - Page 161 – *Usborne Book of Famous Lives*

General Reference Books

- ❑ *Chasing the Moon to China* by Virginia Overton McLean
- ❑ *Count Your Way Through China* by Jim Haskins
- ❑ *Country Insights: China* by Julia Waterlow
- ❑ *Wonders of China* by Lynn M. Stone
- ❑ *The Provinces & Cities of China* by Lynn M. Stone
- ❑ *A Family in China* by Nance Lui Fyson
- ❑ *A Family in Hong Kong* by Peter Otto Jacobsen
- ❑ *Chinese New Year* by Lola M. Schaefer
- ❑ Pages 48 & 49, *Children Just Like Me* by Barnabas & Anabel Kindersley

Literature

- ❑ Chapter 3 - *Missionary Stories with the Millers*
- ❑ *Story About Ping* by Marjorie Flack
- ❑ Ivory Wand – *Stories From Around the World* by Usborne
- ❑ Secret Weapon – Jim Weiss tape
- ❑ *Granny Han's Breakfast* by Sheila Groves
- ❑ *Little Pear* by Eleanor Frances Lattimore
- ❑ *A Little Tiger in the Chinese Night* by Song Nan Zhang
- ❑ *Tikki Tikki Tembo* by Arlene Mosel
- ❑ *The Chinese Mirror* by M. Ginsburg
- ❑ *Li Lun, Lad of Courage* by Carolyn Treffinger
- ❑ *In the Sun* by Huy Voun Lee
- ❑ *In the Snow* by Huy Voun Lee
- ❑ *In the Park* by Huy Voun Lee
- ❑ *At the Beach* by Huy Voun Lee

Science

- ❑ Ducks
  - Duckling - *Baby Animal Stories*
  - *A Duckling is Born* by Hans-Heinrich Isenbart
  - *Ducks Don't Get Wet* by Augusta Goldin
  - http://encarta.msn.com/find/MediaMax.asp?pg=3&ti=761565757&idx=461514740

  Notebook Suggestions:
  1. What helps ducks move through water?
  2. How do ducks get their food?
  3. What is down?

- ❑ Pandas
  - Panda Baby - *Baby Animal Stories*
  - Pages 28 - *Special Wonders of the Wild Kingdom*
  - *Giant Pandas* by Marcia S. Freeman
  - *Wild Bears!* Panda by Tom & Pat Leeson
  - http://encarta.msn.com/find/MediaMax.asp?pg=3&ti=7615797&idx=461517317 (membership required)

  Notebook Suggestions:
  1. Learn what pandas spend most of their day doing.
  2. How do pandas use the thumb-like toe on their front feet?

❑ Sloth Bear

- www.bearbiology.com/sldesc.html
- www.bears-bears.org/slothbear/index.htm
- www.geobop.com/Mammals/Carnivora/Ursidae/Ursus_ursinus/
- http://encarta.msn.com/find/MediaMax.asp?Pg=3&ti=761572258&idx=461532569 (membership required)

Notebook Suggestions:

1. How many cubs do sloth bears usually have?
2. What is the sloth bear's favorite food?
3. Which part of the sloth bear's face has no hair? Put a picture of a sloth bear in your notebook.

❑ Orangutans

- *Orangutans* (A True Book) by Patricia A. Fink Martin
- *How to Babysit an Orangutan* by Tara Darling & Kathy Darling
- *Tropical Forest Animals* by Elaine Landau
- http://encarta.msn.com/find/MediaMax.asp?pg=3&ti=761555522&idx=461517112

Notebook Suggestions:

1. What does the word "orangutan" mean?
2. What color is an orangutan's hair?
3. What do orangutans build for sleeping?
4. How do orangutans sometimes use leaves?

❑ Snow Leopard

- www.primenet.com/~brendel/snolep.html
- www.halcyon.com/mongolia/snowleopard.html
- http://encarta.msn.com/find/MediaMax.asp?Pg=3&ti=761557586&idx=461562982

Notebook Suggestions:

1. What sound do snow leopards make?
2. Why are snow leopards endangered?

Vocabulary

junk   population   farmer

peasant   chopsticks

Music/Art/Projects

1. Color or make the flag of China. Use gold foil stars on red paper. Add to your notebook.

   The flag of China was adopted in 1949. There are five yellow stars are on bright red background. The large star stands for the Communist party. The smaller stars represent the four classes of society: workers, peasants, soldiers, and students.

2. Color or make a map of China.

3. Make Chinese lanterns.
    - Fold a piece of construction paper in half vertically.
    - Make cuts at one inch intervals stopping 1 1/2 from the open edge.
    - Open paper, make a circle and staple.
    - Cut a 1 1/2 inch wide strip of paper. Staple as a handle.

4. Make Chinese food.
    - *Cooking the Chinese Way* by Ling Yu

Internet Resources

- ❑ www.china-embassy.org
- ❑ www.odci.gov/cia/publications/factbook/geos/ch.html
- ❑ www.chinavista.com/travel/greatwall/greatwall.html
- ❑ http://zinnia.umfacad.maine.edu/~mshea/China/great2.html
- ❑ http://zinnia.umfacad.maine.edu/~mshea/China/shang2.html
- ❑ http://encarta.msn.com/find/MediaList.asp?pg=6&mod=2&ti=761573055

Bible

- ❑ Chapter 8 - *Missionary Stories with the Millers*
    - Matthew 4:19
- ❑ Great Wall of China
    - Genesis 11:1-9
- ❑ *Story About Ping*
    - I Peter 2:13-20
- ❑ Leopard
    - Isaiah 11:6

Flag of China

South Korea

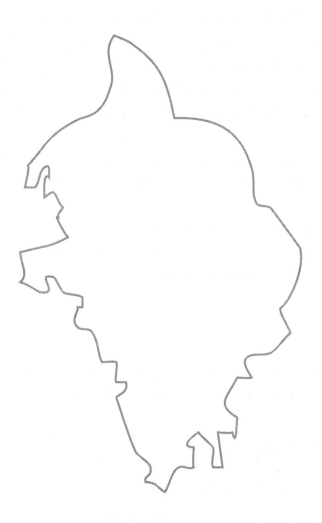

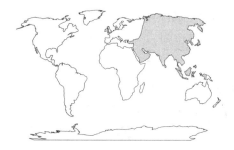

Population_____

Capital City_____

Religion_____

Type of Government_____

Currency_____

Language_____

What are the people called?_____

South Korea

Korea is one of the world's oldest nations. For many years Korea was called the "Hermit Kingdom" because no one knew much about it. During this time, Koreans did not want visitors. They kept very much to themselves. In the late 19th and early 20th centuries, westerners entered Korea as missionaries and traders. At that time, Japan, China, and Russia were fighting for control of northeast Asia. In 1910, Japan was victorious and occupied Korea. Korea remained under Japanese control until the end of World War II. In 1950, North Korea invaded South Korea to unite the country under communism. The Korean War followed. A ceasefire was signed in 1953, but the country is still divided into North and South. The dividing line is known as the 38th parallel.

In Korea, as in many Asian countries, people show respect for each other by bowing. For a very respectful bow a person is on their hands and knees, touching their forehead to their hands.

Korea is about the size of Utah. It is so hilly that some people have said that if the hills could be ironed out it would be larger than Texas. Only 22% of Korea can be farmed. Instead of agriculture, Korea has developed many other industries, such as textiles, chemicals, and electronics.

People/History
- ❑ Bob Pierce
  - • Things That Break God's Heart – chapter 13, *Heaven's Heroes*
  - • Modern Day Elijah chapter 18, *Missionary Stories with the Millers*

General Reference Books
- ☒ *Take a Trip to South Korea* by Keith Lye
- ❑ *Next Stop South Korea* by Fred Martin
- ☒ *Count Your Way Through Korea* by Jim Haskins
- ❑ *A Family in South Korea* by Gwynneth Ashby
- ❑ *Children of the World: South Korea* by Makoto Kubota
- ❑ Pages 54 & 55, *Children Just Like Me* by Barnabas & Anabel Kindersley

Science
- ❑ Lesson 15 – *Considering God's Creation* - The Animal Kingdom
  - • The Wonders of God's Creation: Animal Kingdom – Moody Institute of Science Video

Vocabulary

peninsula	kimchi	pottery
ceramics	palace	bicycle
38th parallel	temple	squid

Music/Art/Projects

1.   Color or make the flag of Korea.

A red and blue circle showing the Chinese symbol of yin and yang, which represents the harmony of opposites in nature, is placed in the center of a white field. The white field represents peace and the white clothing traditionally worn by the Korean people. In each corner of the flag is a set of three black bars, representing heaven, water, fire, and earth.

2.   Color or make a map of Korea.

3.   Make Korean Barbecued Beef.

1 lb. Sirloin steak	1/4 cup soy sauce
3 tablespoons sugar	2 tablespoons oil
1/4 teaspoon pepper	3 green onions, finely chopped
2 cloves garlic, chopped	

Trim fat. Cut beef across the grain in 1/8-inch strips. Mix remaining ingredients; stir in beef until well coated. Refrigerate 30 minutes. Drain beef; stir fry in skillet or wok over medium heat until light brown. Serve with hot cooked rice.

4.   Play "rock, scissors, paper". It originated in Korea where it is called "kawi, bawi, bo".

Internet Resources

❑   http://encarta.msn.com/find/MediaList.asp?pg=6&mod=2&ti=761562354

❑   www.lifeinkorea.com/pictures

❑   www.koreafolkart.com

❑   www.quaypress.com/korea

Bible

❑   Chapter 18 - *Missionary Stories with the Millers*

• 1 Kings 17:1

• James 5:17

Flag of South Korea

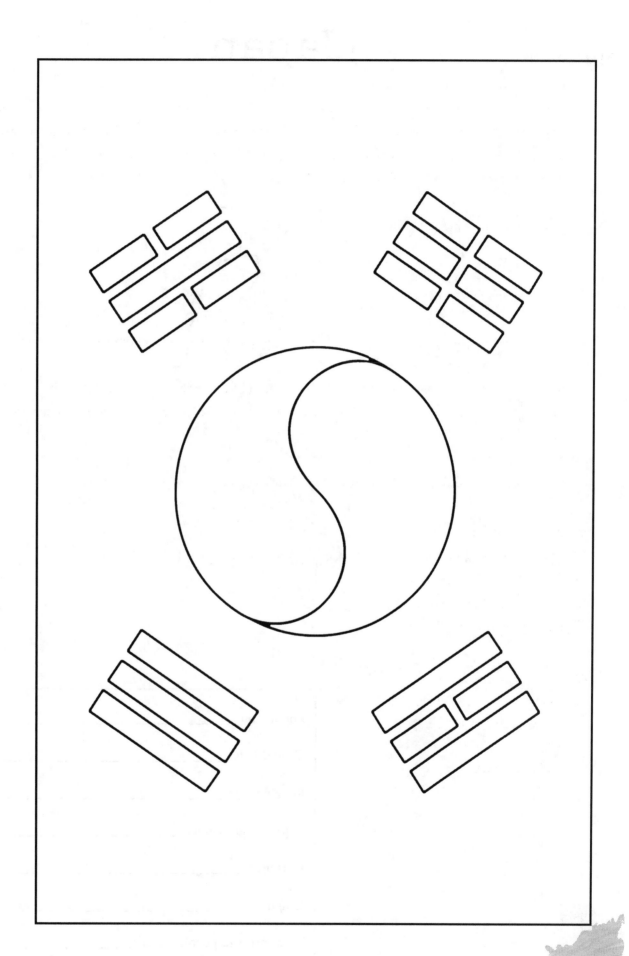

Japan

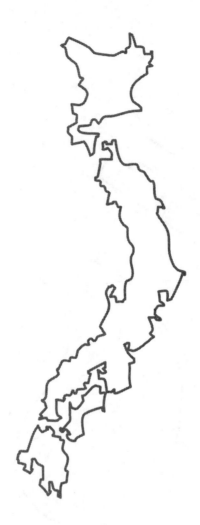

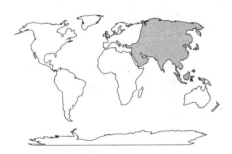

Population_____

Capital City_____

Religion_____

Type of Government_____

Currency_____

Language_____

What are the people called?_____

Japan

Japan is one of the most densely populated countries in the world. Most people live on the narrow coastal plains. Over three-fourths of the population is crowded into towns and cities. All together, Japan is no bigger than the state of Montana. In this small space resides a population half the size of the United States. Japan's capital city, Tokyo, and its main seaports and industrial cities are on the island of Honshu. Most of the population of Japan lives on Honshu.

Japan is a beautiful country. There are many high mountains with rushing rivers and waterfalls. The most beautiful of the mountains is Fujiyama, near Tokyo. It is considered one of the most perfectly formed mountains in the world – its tall, snow-capped peak seems to rise straight out of the ocean. Land for farming is scarce. Every inch of good soil is carefully cultivated, and even the sides of mountains are farmed. Even though the farmers use every bit of land they can, they still cannot grow enough to feed the whole nation. Japan depends on industry. Japan sells its goods to other countries to earn enough money to buy the food it needs.

Most homes in Japan are small since the land is expensive and crowded. The homes are simply furnished. The straw mats on the floor are called tatami. Bedding is a futon. Most Japanese homes have a traditional nook called a tokonoma (corner of beauty). In the tokonoma, a painted scroll hangs on the wall above a platform or low table holding a graceful vase of flowers. The Japanese have strong family ties and deep respect for authority. In Japan, it is polite to greet each other by bowing. Ikebana, the art of flower arranging, is practiced throughout Japan. Other art forms are: the cultivation of miniature bonsai trees, the writing of haiku and tanka poetry, and origami, the art of paper folding.

Cleanliness is such an important part of Japanese life that even taking a bath becomes a ritual. First a deep oval-shaped wooden or porcelain tub is filled with very hot water. The actual washing is done outside the tub! Sitting on a low stool and using a wooden bowl, water is splashed all over the body. Don't worry about the mess there is a drain in the floor. Only after the body is soaped and rinsed thoroughly do they get in the tub. Since many houses do not have bathtubs, there are public bathhouses with tubs large enough to accommodate forty people at once. Because bathing is also a social occasion, much chattering can be heard from the separate men's and women's halls.

General Reference Books

- ☐ *Land of the Rising Sun* – Countries and Their Children
- ☐ *A Visit to Japan* by Peter & Connie Roop
- ☐ *A Family in Japan* by Peter Otto Jacobsen
- ☐ *Japan* (A True Book) by Ann Heinrichs
- ☐ *Count Your Way Though Japan* by Jim Haskins
- ☐ *Look What Came From Japan* by Miles Harvey
- ☐ *Japan* by Tamara L. Britton
- ☐ *Next Stop Japan* by Clare Boast
- ☐ Pages 52 & 53, *Children Just Like Me* by Barnabas & Anabel Kindersley

Literature

- ☐ *Grandfather's Journey* by Allen Say
- ☐ *A Carp for Kimiko* by Virginia Kroll
- ☐ *The Bicycle Man* by Allen Say
- ☐ *The Wise Old Woman* by Yoshiko Uchida
- ☐ *The Two Foolish Cats* by Yoshiko Uchida
- ☐ *The Rooster Who Understood Japanese* by Yoshiko Uchida
- ☐ *How My Parents Learned to Eat* by Ina Friedman
- ☐ *Crow Boy* by Taro Yashima
- ☐ *Japanese Children's Favorite Stories* edited by Florence Sakade
- ☐ Two Monks – Jim Weiss tape

Science

- ☐ Lesson 19 – *Considering God's Creation* - Introduction to Mammals

Vocabulary

futon	bonsai	kimono
yen	island	tatami

Music/Art/Projects

1. Color or make the flag of Japan.

   Trace a CD on a plain piece of paper and let the child color it. Or, trace a CD on a piece of red paper, let the child cut out the circle and glue it on a piece of plain paper. The red circle, or "rising sun", of the Japanese flag stands for warmth, brightness, and sincerity. The white background symbolizes purity and integrity. The flag was adopted in 1854.

2. Color or make a map of Japan.

3. Make Origami.

   Check your local library for books on origami. Craft and teacher supply stores have brightly colored paper perfect for origami.

4. Play Karuta.

   Poetry is so much a part of everyday life in Japan that a card game called Karuta, or "One Hundred Poems", is very popular.
   - Copy half of popular or familiar poems onto two index cards. First half on one card; second half on second card. Keep the first half in a separate stack.
   - Choose a leader to read the opening half of a well-known poem from a card; whoever matches the card with the other half of the poem (among the many spread out on the table) gets to keep it.
   - The player with the most cards at the end of the game is the winner.

5. Have a crab race.

   To play this game, have the children rise up on all fours with their back facing the floor. Have them "race" to the finish line.

6. Make Japanese food.

   Cooking the Japanese Way by Reiko Weston

7. If your budget allows, go to a Japanese restaurant that cooks the food hibachi style where the chef chops and cooks the food at your table.

Internet Resources

❑ www.japan-guide.com

❑ http://encarta.msn.com/find/MediaList.asp?pg=6&mod=2&ti=761566679

Bible

❑ *Grandfather's Journey*
   - Matthew 7:12
   - James 3:18
   - Psalm 34:18
   - 2 Corinthians 1:3-4
   - Psalm 137:1-5

❑ *The Wise Old Woman*
   - Proverbs 8:12
   - Proverbs 10:14
   - Proverbs 17:24
   - Proverbs 12:15
   - Romans 12:3
   - Luke 18:14
   - Philippians 2:3
   - I Peter 5:5

❑ *Two Foolish Cats*
   - Matthew 5:42
   - Philippians 2:4, 20-21
   - Proverbs 14:21
   - Acts 20:35
   - Ecclesiastes 7:9
   - Proverbs 3:35

❑ *The Rooster Who Understood Japanese*
   - Proverbs 19:15
   - Proverbs 13:20
   - Proverbs 11:13
   - Proverbs 25:20-21
   - Luke 6:35

Flag of Japan

India

Religion plays a vital role in the Indian way of life. India's traditions are strongly rooted in religion and greatly influence their music, customs, dance, festivals, and clothing. About 83% of the Indian people are Hindus, and about 11% are Muslims. The next largest religious groups are Christians, Sikhs, Buddhists, and Jains. When many Indians are told of Jesus they readily "accept" Him – they place a picture of Him amongst the other gods that they worship. The biggest challenge to missionaries and other Christians is getting the people to realize that there is only one true God.

The Taj Mahal is one of the world's most magnificent buildings. The white marble monument was built more than 300 years ago by an emperor in memory of his wife. It took over 20,000 workers twenty-one years to build.

Clothing worn by Indians varies greatly by region. Members of the various religious groups may also dress differently. Many men wear a dhoti (a white garment wrapped between the legs). Most women wear a sari (a straight piece of cloth draped around the body as a long dress).

People/History
- [ ] Amy Carmichael
  - Temple Runaway – chapter 20, *Missionary Stories with the Millers*

- [ ] William Carey
  - Boy Who Was Determined – chapter 23, *Missionary Stories with the Millers*
  - *The Shoemaker Who Pioneered Modern Missions* by Ben Alex

- [ ] Gandhi
  - *What's Their Story?: Gandhi* by Pratima Mitchell
  - *Gandhi: Peaceful Warrior* by Rae Bains

General Reference Books
- [ ] *India* by David Cumming
- [ ] *A Family in India* by Tony Tigwell
- [ ] Pages 62-68 *Our Father's World*
- [ ] Pages 56-59, *Children Just Like Me* by Barnabas & Anabel Kindersley

Literature
- [ ] She Changed Her Mind – chapter 13, *Missionary Stories with the Millers*
- [ ] Saved In the Night – chapter 15, *Missionary Stories with the Millers*
- [ ] Snake Charmer – *Stories From Around the World* by Usborne
- [ ] *To the Top* by S.A. Kramer
- [ ] *Little Black Sambo* by Helen Bannerman
- [ ] *The Story of Little Babaji* by Helen Bannerman
- [ ] *Adventures of Mohan* (Rod & Staff)

India

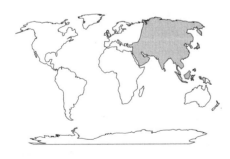

36

Population_____

Capital City_____

Religion_____

Type of Government_____

Currency_____

Language_____

What are the people called?_____

Science

When studying wildlife: 1) Read a book on the species. 2) In notebook describe or draw a picture of the species, describe their habitat (where they live and what they eat), and tell what makes the species unique or use the Animal Report form on page 235. If possible, use a scripture verse that mentions the species.

❑ Elephants
  - Lesson 20 – *Considering God's Creation*
  - Page 22 & 23 – *Special Wonders of the Wild Kingdom*
  - *Elephants* by Louise Martin
  - www.pbs.org/edens/bhutan/a_elephant.htm
  - http://encarta.msn.com/find/MediaList.asp?pg=6&mod=2&ti=761575386
  - *Little Big Ears* by Cynthia Moss (This is a wonderful story of a baby elephants first year of life. Since it is set in Africa, you may use it here or in the study of Africa.)

  Notebook Suggestions:
  1. What is the name for a female elephant?
  2. What do elephants use as hands?
  3. What do elephants eat?
  4. How do elephants use their trunks?
  5. How can you tell an African elephant from an Indian elephant?
  6. Where do elephants live?
  7. What eats elephants?
  8. What is the biggest threat to elephants?

❑ Tigers
  - Tiny Tiger - *Baby Animal Stories*
  - Pages 62 & 63 – *Special Wonders of the Wild Kingdom*
  - *Tigers* by Louise Martin
  - www.pbs.org/edens/bhutan/a_tiger.htm
  - http://encarta.msn.com/find/MediaList.asp?pg=6&mod=2&ti=761576290

  Notebook Suggestions:
  1. The tiger is the largest member of which family?
  2. Tigers are found only in _____ .
  3. What animals do tigers hunt?
  4. What do tigresses teach their cubs?

❑ Water Buffalo
  - www.animalinfo.org/species/artiperi/bubaarne.htm

  Notebook Suggestions:
  1. What do water buffalo like to rest in?
  2. Why do water buffalos roll in the mud?
  3. How do water buffalo help with rice farming?

❑ Chevrotain
  • www.geobop.com/Mammals/Artiodactyla/Tragulidae/
  • www.channel4000.com/partners/mnzoo/chevrotain.html
  • www.encyclopedia.com/searchpool.asp?target=@DOCTITLE%20chevrotain

Notebook Suggestions:

1. How do chevrotains defend themselves?
2. Where do chevrotains live?
3. What do chevrotains eat?
4.) Chevrotains are also called _____ .

Vocabulary

poachers	caste system	curry
untouchables	laborer	hindi
bazaar	bullock	henna plant
village	gem	

Internet Resources

❑ http://encarta.msn.com/find/MediaList.asp?pg=6&mod=2&ti=761557562
❑ www.uni-giessen.de/~gk1415/india.htm
❑ www.harappa.com
❑ www.indonet.com/HolidaysandFestivals.html

Music/Art/Projects

1. Color or make the flag of India.

   India's flag has three equal horizontal bands: deep saffron (for courage and sacrifice) at the top, white (for purity and truth) in the middle, and dark green (for faith and fertility) at the bottom. The wheel in the center is an ancient symbol called the Dharma Chakra (Wheel of Law).

2. Color or make a map of India.

3. Make Indian Flat Bread.

2 cups all-purpose flour	1/4 cup plain yogurt
1 egg, slightly beaten	1 1/2 teaspoons baking powder
1 teaspoon sugar	1/4 teaspoon salt
1/8 teaspoon baking soda	1/2 cup milk
oil	poppy seeds

   Mix all ingredients except milk, oil, and poppy seeds. Stir in enough milk to make a soft dough. Turn dough onto lightly floured surface. Knead until smooth, about 5 minutes. Place in greased bowl; turn greased side up. Cover, let rise in warm place 3 hours. Divide dough into 6 or 8 equal parts. Flatten each part on lightly floured surface, rolling it out to 6"x4", 1/4 inch thick. Brush with oil; sprinkle with poppy seeds. Place 2 cookie sheets in oven; heat oven to 450° F. Remove cookie sheets from oven; place breads on hot cookie sheets. Bake 6-8 minutes.

4.   Play Parcheesi.

Bible

❑ Chapter 15 - *Missionary Stories with the Millers*
   • Luke 8:22-25
❑ Chapter 20 - *Missionary Stories with the Millers*
   • Isaiah 61:1
   • Luke 4:18
❑ Chapter 23 - *Missionary Stories with the Millers*
   • Isaiah 54:6a
   • Matthew 28:19

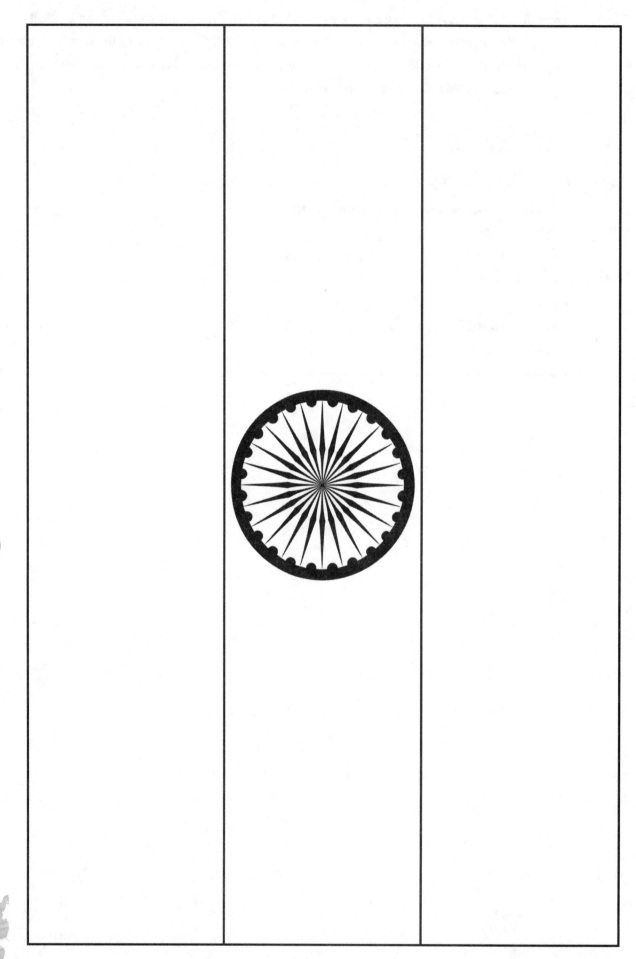

Flag of India

40

connect the dots

Start Here →

1
2
3
4
5
6
7
9
10
11
12
13
14
15
16
17
18
19
20
21
22
23
24
25
26
27
28
29
30
31
32
33
34
35
36
37
38

41

FIND THE TWINS

Which two are exactly alike?

1

2

3

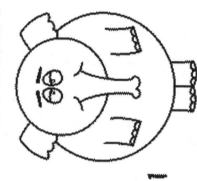

4

5

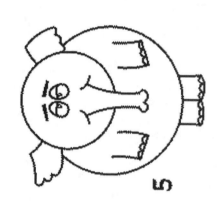

6

42

Israel

On May 14, 1948, Israel was established as a Jewish state. Historically, it is considered the Holy Land for Jews, Christians, and Muslims. Most of the people of Israel are Jews, about half of whom were born in other countries. Even Jews who live elsewhere consider Israel their spiritual home. Israel makes up most of the region once known as Palestine. Jerusalem is considered the holiest of all cities. No other city in the world is so important to such a great number of people with such different beliefs. A favorite pastime of Israelis is reading. On average, more books are read and published in Israel than in any other country. Israel also has more newspapers per capita than any other country. The Biblical Zoo in Jerusalem is a favorite place for children in Israel. There you will find every living creature that is mentioned in the Bible.

Israel's Dead Sea is three times saltier than ocean water. The water is so thick with salt and minerals that everything floats on the surface.

People/History
- ☐ Jesus
- ☐ Abraham
- ☐ Isaac
- ☐ Jacob
- ☐ Children of Israel as God's chosen people

General Reference Books
- ☐ *Passport to Israel* by Clive Lawton
- ☐ *We Live in Israel* by Gemma Levine
- ☐ *Take a Trip to Israel* by Jonathan Rutland
- ☐ *A Kibbutz in Israel* by Allegra Taylor
- ☐ *Count Your Way Through Israel* by James Haskins
- ☐ *A Kid's Catalog of Israel* by Chaya M. Burstein
- ☐ *Hanukkah* by Lola M. Schaefer
- ☐ *Dance, Sing, Remember* by Leslie Kimmelman
- ☐ Pages 60 & 61, *Children Just Like Me* by Barnabas & Anabel Kindersley

Literature
- ☐ *Mrs. Katz and Tush* by Patricia Polacco
- ☐ *Matzah Ball* by Mindy Avra Portnoy
- ☐ *Sammy Spider's First Shabbat* by Sylvia A. Rouss
- ☐ *Sammy Spider's First Passover* by Sylvia A. Rouss
- ☐ *Sammy Spider's First Hanukkah* by Sylvia A. Rouss
- ☐ *Joseph Who Loved the Sabbath* by Marilyn Hirsh

Israel

The Middle East

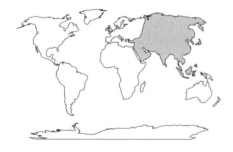

Population_____

Capital City_____

Religion_____

Type of Government_____

Currency_____

Language_____

What are the people called?_____

44

Science

❑ Donkey
   - Farm Donkey - *Baby Animal Stories*
   - http://encarta.msn.com/find/MediaMax.asp?pg=3&ti=761556565&idx=461539067

   Notebook Suggestions:
   1. Explain how donkeys are different from horses.
   2. What do you call the sound a donkey makes?

Vocabulary

shalom	menorah	seder	kosher
ibbutz	shofar	Torah	

Music/Art/Projects

1. Color or make the flag of Israel.
    The blue and white colors of the flag are taken from the Jewish prayer shawl. The central emblem is the Star of David, which has long been associated with the Jewish people.

2. Color or make a map of Israel.

3. Listen to Jewish music. Check your local library for selections.

4. Make Jewish food. *Cooking the Israeli Way* by Josephine Levi-Bacon

5. *Jewish Holiday Games for Little Hands* by Ruth Esrig Brinn

Internet Resources

❑ www.israel.org
❑ http://encarta.msn.com/find/MediaList.asp?pg=6&mod=2&ti=761575008

Bible

❑ The story of Abraham
   - Genesis 12 & 13
   - Genesis 15
   - Genesis 17 & 18
   - Genesis 21 & 22

❑ The story of Isaac
   - Genesis 21:1-5
   - Genesis 24
   - Genesis 25:19-28:5

❑ The story of Jacob
   - Genesis 25:19-34
   - Genesis 27-35
   - Genesis 46-50

❑ *Mrs. Katz and Tush*
   - Leviticus 19:18
   - Luke 10:27
   - Matthew 22:39
   - Matthew 7:12
   - Luke 6:31
   - Psalm 87

❑ Donkey
   - Matthew 21:1-7
   - Proverbs 26:3
   - Genesis 22:3
   - Job 39:5-8
   - Job 6:5

Flag of Israel

Word Search

```
i  s  r  a  e  l  d  a  u
j  u  p  x  n  y  n  s  q
k  o  r  e  a  j  p  i  e
k  n  m  a  e  e  c  a  q
e  k  y  g  a  v  h  n  c
x  z  j  a  p  a  n  v  h
o  o  t  b  h  n  m  r  i
n  q  v  j  c  a  u  t  n
i  n  d  i  a  o  q  h  a
```

Asia   China
India   Israel
Japan   Korea

Asia Review Map

See how many countries you can identify. Write their names on the map.

ad maiorem Dei glo

 **Notes:**

Europe

Europe is the smallest inhabited continent, but it has the third largest population. Europe is a very diverse continent with many different cultures. The terrain varies from vast, rich farmland to splendid, towering mountains. The highest point in Europe is in Russia at Mt El'brus (18,510 ft), and the lowerst point is in the Caspian Sea (92 ft. below sea level). Russia is located on Europe and Asia. For this study, it is covered in Europe

Map of Europe

Color each country you study.

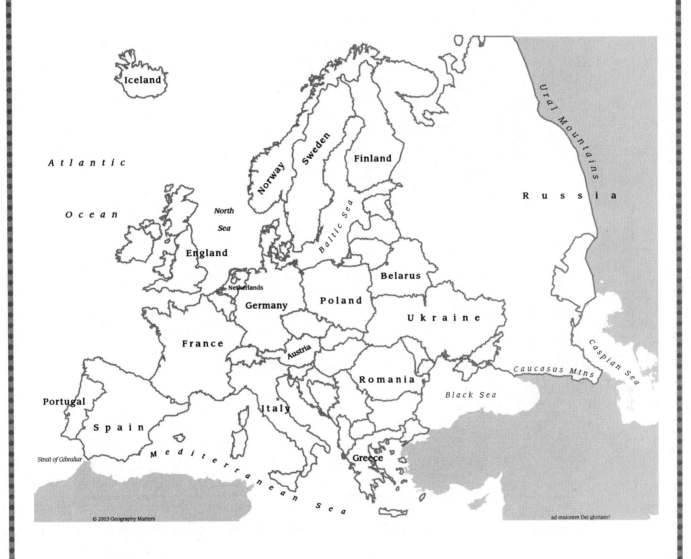

Iceland

Atlantic

Ocean

North
Sea

Norway

Sweden

Finland

Baltic Sea

Ural Mountains

R u s s i a

England

Netherlands

Germany

Belarus

Poland

U k r a i n e

France

Austria

Caspian Sea

Caucasus Mtns

Portugal

Romania

Black Sea

Italy

S p a i n

Greece

Strait of Gibraltar

M e d i t e r r a n e a n   S e a

© 2003 Geography Matters

ad maiorem Dei gloriam!

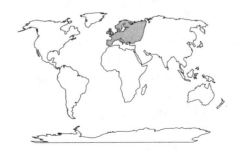

Russia

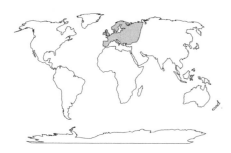

Population_____

Capital City_____

Religion_____

Type of Government_____

Currency_____

Language_____

What are the people called?_____

Russia

Most of Russia's people are ethnic Russians, meaning they are descended from Slavic people. More than 100 minorities also live in Russia. The Russian people are used to hardship and change, but they are strong of spirit and warm of heart. Russian children celebrate their birthday twice a year. Once on the day they were born and again on the day of the saint they were named after. Russians have birthday pie instead of birthday cake. When a child loses a tooth in Russia, they put it under their pillow for the Tooth Mouse. If the tooth is gone in the morning, a new tooth is sure to replace the old one.

Russia has many natural resources. Vast deposits of petroleum, natural gas, coal, gold, and iron ore are a few. Many of these reserves lie far from settled areas. Russia's severe cold climate makes it difficult to take advantage of the country's many valuable resources.

People/History

❑ Catherine the Great
- Page 197 – *Usborne Book of Famous Lives*

❑ Peter Tchaikovsky
- Tchaikovsky Discovers America – Classical Kids audio series
- The Story of Tchaikovsky – Music Masters Series
- *Getting to Know the World's Greatest Composers: Tchaikovsky* by Mike Venezia
- *The Nutcracker* by Carin Dewhirst

❑ Peter the Great
- *Peter the Great* by Diane Stanley

General Reference Books

❑ *Next Stop Russia* by Clare Boast
❑ *Cities of the World: St. Petersburg* by Deborah Kent
❑ *Eyewitness Books: Russia* by Kathleen Berton Murrell
❑ *Russia* by Kristin Thoennes
❑ *Look What Came From Russia* by Miles Harvey
❑ *Russia* by Bob Italia
❑ *Russian Federation* (True Book) by Karen Jacobsen
❑ Pages 30 & 31, *Children Just Like Me* by Barnabas & Anabel Kindersley

Literature

❑ *Another Celebrated Dancing Bear* by Gladys Scheffrin-Falk
❑ *The Peddler's Gift* by Maxine Rose Schur
❑ *The Fool of the World and the Flying Ship* A Russian Folktale
❑ *The Cat and the Cook* by Ethel Heins

- ❑ *Good Morning Chick* by M. Ginsburg
- ❑ *My Mother is the Most Beautiful Woman in the World* by Becky Reyher
- ❑ *The Mitten* by Jan Brett
- ❑ *The Gossipy Wife* by Amanda Hall
- ❑ *One Fine Day* by Hogrogrian

Science

- ❑ Lesson 22 – *Considering God's Creation* - Animal Structure

- ❑ Bears
  - Page 34 - *Special Wonders of the Wild Kingdom*
  - *Baby Grizzly* by Beth Spanijian
  - Bear Cub - *Baby Animal Stories*
  - *Grizzly Bear* by Berniece Freschet
  - *Bear* by Mike Down
  - *Bears* (A New True Book) by Mark Rosenthal
  - http://encarta.msn.com/find/MediaList.asp?pg=6&mod=2&ti=761572258

  Notebook Suggestions:
  1. Which senses are strongest in bears?
  2. What do omnivores eat?
  3. What is hibernation?
  4. How do bears get ready for hibernation?

- ❑ Wolves
  - *Watchful Wolves* by Ruth Berman
  - *Wolves* by Karen Dudley
  - *Look to the North: A Wolf Pup Diary* by Jean Craighead George
  - *Reading About the Gray Wolf* by Carol Greene
  - http://encarta.msn.com/find/MediaMax.asp?pg=3&ti=761560395&idx=461546834

  Notebook Suggestions:
  1. What do wolves hunt?
  2. Which wolf is the leader of the pac?.
  3. How do wolves show respect to the leader?
  4. Why do wolves pant?

Vocabulary

czar	communist	atheist
ruble	ornate	citadel
babushka	ballet	icon
	tutu	performance

54

Music/Art/Projects

1.   Go to the ballet or rent a video of *The Nutcracker*.

2.   Listen to Tchaikovsky.

3.   Color or make the flag of Russia.
      Adopted in 1991, the Russian flag is a tricolor; three equal bands of white, blue, and red. It had been the unofficial ethnic flag of the Russian people since 1988.

4.   Color or make a map of Russia.

5.   Make pretzels.

1 envelope dry yeast	1 1/2 cups lukewarm water
3/4 teaspoon salt	1 1/2 teaspoons sugar
4 cups flour	1 egg, beaten
coarse (kosher) salt	

      Directions: In a large bowl dissolve yeast in water. Add sugar and salt. Mix in flour and knead until the dough is soft and smooth. Do not let dough rise. Divide immediately into smaller pieces and roll into ropes. Form the ropes into circles or pretzel shapes. Place on a cookie sheet covered with foil and dusted with flour. Brush each pretzel with the beaten egg mixed with a little water and sprinkle with coarse salt. Bake in a 400° oven until brown.

6.   Make Russian food.
      - *Cooking the Russian Way* by Gregory Plotkin
      - *Russian Food and Drink* by Valentina Lapenkova

Internet Resources

❑ www.interknowledge.com/russia/

❑ http://encarta.msn.com/find/MediaList.asp?pg=6&mod=2&ti=761569000

Bible

❑ *Another Celebrated Dancing Bear*
   - Proverbs 17:17
   - John 15:12-13
   - 2 Samuel 17:8
   - 1 Samuel 17:34-37
   - 2 Kings 2:23-25

❑ *The Mitten*
   - Isaiah 1:18

❑ *One Fine Day*
   - Genesis 47:16-19

❑ Bears
   - Isaiah 11:7
   - Isaiah 59:11

❑ Wolf
   - Genesis 49:27
   - John 10:7-15
   - Zephaniah 3:1-5
   - Matthew 7:15-20
   - Isaiah 11:6

Flag of Russia

Great Britain

Great Britain is an island nation just off the coast of Europe. England, Scotland, and Wales form Great Britain. The United Kingdom brings together England, Scotland, the principality of Wales and the province of Northern Ireland. The Isle of Man and the Channel Islands are self-ruling states associated with the United Kingdom. The British Isles includes about 5,000 small islands and two large ones, Great Britain and Ireland. London, England's capital, is home to about 6.8 million people.

England's traditional industries of coal, textiles, and shipbuilding have all declined in recent years. The English economy today is based upon finances, services, chemicals, and telecommunications. Wales is sheep-farming country with three times as many sheep as people. The industrialized south now depends on foreign-owned factories rather than its coalfields. Scotland's most important resource is oil, produced offshore in the North Sea. Important exports include salmon and whiskey.

The province of Northern Ireland has seen long years of conflict between those who wish to remain with the United Kingdom and those who wish to be part of the Republic of Ireland. A peace agreement was reached in 1998.

People/History

❑ William Shakespeare
- *Bard of Avon* by Diane Stanley
- *What's Their Story? William Shakespeare* by Haydn Middleton

❑ Elizabeth I
- Pages 162 & 185 – *Usborne Book of Famous Lives*
- *Young Queen Elizabeth* by Francine Sabin
- *Good Queen Bess* by Diane Stanley

❑ Charles Dickens
- *The Man Who Had Great Expectations* by Diane Stanley

❑ Beatrix Potter
- *Beatrix Potter* by John Malam
- *Country Artist* by David R. Collins

❑ Florence Nightingale
- Page 214 – *Usborne Book of Famous Lives*
- *Florence Nightingale* – Young Christian Library
- Florence Nightingale – Animated Hero Classics Video
- The Angel of the Crimea – Your Story Hour audio series
- *A Picture Book of Florence Nightingale* by David Adler

Great Britain

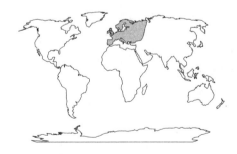

Population_____

Capital City_____

Religion_____

Type of Government_____

Currency_____

Language_____

What are the people called?_____

General Reference Books

- ❑ *Roundabout the British Isles* – Countries and Their Children
- ❑ *A Family in England* by Jetty St. John
- ❑ *England* (A True Book) by Michael Burgan
- ❑ *Take a Trip to England* by Chris Fairclough
- ❑ *Passport to Great Britain* by Nicola Wright
- ❑ *We Live in Britain* by Chris Fairclough
- ❑ *A Taste of Britain* by Roz Denny
- ❑ *Cooking the English Way* by Barbara W. Hill

Literature

- ❑ *Tales From Shakespeare* by Charles and Mary Lamb
- ❑ Shakespeare for Children – audio by Jim Weiss
- ❑ *Young Persons Guide to Shakespeare* by Anita Ganeri
- ❑ *Tale of Peter Rabbit* by Beatrix Potter
- ❑ *Mr. Gumpy's Motor Car* by John Burningham
- ❑ *Mr. Gumpy's Outing* by John Burningham
- ❑ My Heart's in the Highlands – poem by Robert Burns
- ❑ *Winnie the Pooh* by A. A. Milne
- ❑ *Robin Hood* retold by Margaret Early
- ❑ *St. George & the Dragon* by Margaret Hodges
- ❑ Dick Whittington – *Stories From Around The World* by Usborne
- ❑ *Mary Jones & Her Bible* by Mary Ropes

Science

- ❑ Sheep
  - *Baby Lamb* by Beth Spanjian
  - *Smudge, the Little Lost Lamb* by James Herriot
  - *The Sheep Book* by Dorothy Hinshaw Patent
  - *Sheep* (A True Book) by Sara Swan Miller
  - Pages 25-30 – *Farm Animals* (A New True Book) by Karen Jacobsen
  - http://encarta.msn.com/find/MediaMax.asp?pg=3&ti=761559678&idx=461514729

  Notebook Suggestions:
  1. For what are sheep raised?
  2. List uses for wool.

Vocabulary

moor	tea	royalty
thistle	monarchy	thatch
textile	bagpipe	kilt
tartan		

Although English is spoken in both the United States and Great Britain, not all words are the same.

bloke – guy	trainers - sneakers
mate – friend	ice lolly - popsicle
petrol – gasoline	telly - television
biscuit – cookie	jumper - sweater
boot – car trunk	bonnet - hood
lift – elevator	fringe - bangs
tights – panty hose	bangers - sausages
I'll ring you – I'll call you	bobby - police officer

Music/Art/Projects

1. Color or make the flag of Great Britain.

   The British flag is commonly referred to as the "Union Jack". The white saltire on a blue field was taken from the St. Andrew's Cross of Scotland. The diagonal red saltire is from Ireland's St. Patrick. The central red with white cross was adapted from England's St. George's Cross.

2. Color or make a map of Great Britain.

3. Listen to a recording of bagpipes.
   Check your local library.

4. Go to tea at a local teahouse or have a tea party at home.

5. Make Cheddar Cheese Soup.

1 small onion, chopped	1 stalk celery, sliced thin
2 tablespoons butter	2 tablespoons flour
1/4 teaspoon pepper	1/4 teaspoon dry mustard
10 3/4 oz. condensed chicken broth	1 cup milk
2 cups shredded cheddar cheese	paprika

   Cover and simmer onion and celery in margarine in 2-quart saucepan until onion is tender, about 5 minutes. Stir in flour, pepper, and mustard. Cook over low heat, stirring constantly until smooth and bubbly; remove from heat. Add chicken broth and milk. Heat to boiling over medium heat, stirring constantly. Boil and stir 1 minute. Stir in cheese; heat over low heat, just until cheese is melted. Sprinkle soup with paprika.

Internet Resources

- ❑ www.royal.gov.uk
- ❑ www.number-10.gov.uk
- ❑ www.castles-of-britain.com
- ❑ www.britannia.com/history/h6.html
- ❑ www.odci.gov/cia/publications/factbook/geos/uk.html
- ❑ http://encarta.msn.com/find/MediaList.asp?pg=6&mod=2&ti=761553483

Bible

- ❑ *The Tale of Peter Rabbit*
  - Matthew 26:36
  - John 18:1
  - John 19:41
  - Luke 23:43
  - 2 Timothy 2:6
  - James 5:7
  - Ephesians 6:1
- ❑ Sheep
  - Matthew 25:31-33
  - Luke 2:8-20
  - Exodus 2:16-19
  - Genesis 4:2-5
  - Daniel 8:3-8
  - 2 Kings 3:4
  - John 1:29
  - Matthew 10:16
  - Ezekiel 34

Flag of Great Britain

Crossword Puzzle

Across
2 Famous forest.
3 River in 1 down.
4 Famous clock in 1 down.
5 Word used for gasoline.
6 Word used for friend.
8 The name of the Queen.

Down
1 The capital of England.
2 Famous playwrite.
7 Animal raised for its wool.

France

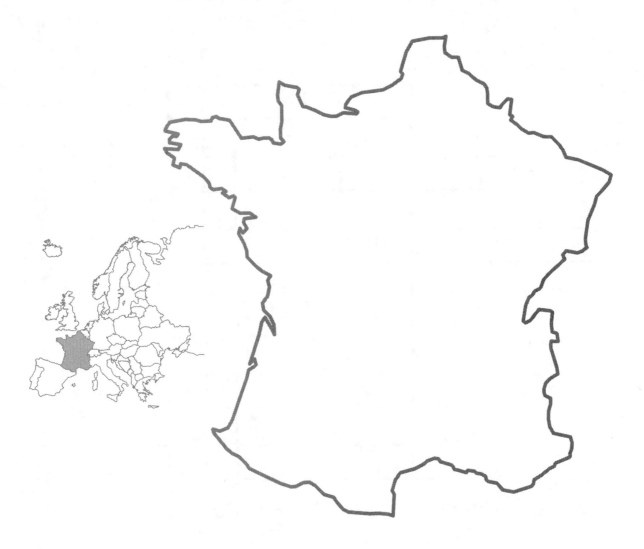

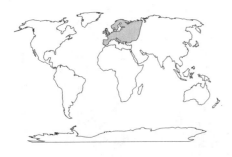

Population_____

Capital City_____

Religion_____

Type of Government_____

Currency_____

Language_____

What are the people called?_____

France

France is the largest country in Western Europe. Often called the "gateway to Europe", it is rich in history and art. A country smaller than Texas, its snowcapped Alps provide a skier's paradise and its Riviera, a sunny vacation spot. While its many rivers and ideal climate provide farmers fertile lands to raise wheat, vegetables and grapes for wine, its flowers are made into elegant French perfumes.

The Eiffel Tower is one of Paris' most famous landmarks. It was built by Gustave Eiffel in 1889 for the International Exposition. On a clear day it is possible to see over 40 miles from the top of the tower.

French is a beautiful language that is flowing and harmonious to hear. Over the years, the French have borrowed many words from English. Some people feared that the French language would be corrupted and weakened by the use of too many foreign words. A law now forbids the official use of English words where there is already a French word in existence.

To the French, cooking is an art and greatly adds to the enjoyment of life. They take great pride and pleasure in cooking. Their chefs and cuisine are world famous. The French believe that good food deserves fresh ingredients, time, and attention. It is not only important to cook well in France, but to shop well, too.

People/History

❑ Marie Curie
- Pages 85 & 231 – *Usborne Book of Famous Lives*
- *Marie Curie: Brave Scientist* by Keith Brandt
- *Marie Curie's Search For Radium* by Beverley Birch & Christian Birmingham

❑ Louis Pasteur
- Page 82 – *Usborne Book of Famous Lives*
- *Louis Pasteur Young Scientist* by Francene Sabin
- *Pasteur's Fight Against Microbes* by Beverley Birch & Christian Birmingham
- Solving the Riddle – Your Story Hour audio series

❑ Claude Monet
- *Getting to Know the World's Greatest Artists: Monet* by Mike Venezia
- *Monet* by Jude Welton (Eyewitness Books by Dorling Kindersly)
- *Art for Young People: Claude Monet* by Peter Harrison
- *Monet* by Vanessa Potts – A wonderful book with a photograph of Monet's work on one page and a short discussion or background on the work on the opposite page.

❑ Joan of Arc
- *Joan of Arc* by Diane Stanley
- *Joan of Arc* by Margaret Hodges
- The Girl General – Your Story Hour audio series

General Reference Books

- *Usborne First Book of France*
- *Cooking the French Way* by Lynne Marie Waldee
- *France* by Michael Dahl
- *Picture a Country: France*
- *Look What Came From France* by Miles Harvey
- *Count Your Way Through France* by Jim Haskins
- *A Family in France*
- Page 32, *Children Just Like Me* by Barnabas & Anabel Kindersley

Literature

- *Giraffe That Walked to Paris* by Nancy Milton
- *Bon Appetit, Bertie!* by Joan Knight
- *Mirette on the High Wire* by Emily Arnold McCully
- *Marie in Fourth Position* by Amy Littlesugar
- *Madeline* by Ludwig Bemelmans
- *Puss in Boots* retold by Lorinda Bryan Cauley
- *Little Red Riding Hood* retold by James Marshall
- Three Musketeers audio recording by Jim Weiss
- *New Coat for Anna* by Harriet Ziefert
- *Linnea in Monet's Garden* by Christina Bjork
- *Glorious Flight* by Alice and Martin Provensen

Science

- Trees
  - *Usborne 1st Nature Trees*
  - *Ultimate Trees & Flowers Sticker Book*
  - *Trees* by Linda Gamlin
  - *Oak Trees* by Marcia S. Freeman
  - *How a Seed Grows* by Helene J. Jordan
  - *Leaves* by Gail Saunders-Smith
  - *Why Do Leaves Change Color* by Betsy Maestro
  - *The Secret Life of Trees* by Chiara Chevallier
  - *Discovering Trees* by Keith Brandt
  - Lesson 11 *Considering God's Creation*

Notebook Suggestions:

1. Collect pictures of trees, and give a description of each tree.
2. Note which of these trees are mentioned in the Bible and give Scripture reference.
3. Collect or draw a picture of a leaf from these trees.
4. Name four trees used primarily for lumber.

5. Name five trees that bear food.

6. Name three coniferous trees.

7. Name five deciduous trees.

Vocabulary

coniferous	deciduous	evergreen
trunk	branch	crepe
chateau	vineyard	truffle
cathedral	museum	croissant

Music/Art/Projects

1. Color or make the flag of France.

   Cut red, white, and blue paper evenly into thirds. Let the child glue the strips onto a piece of paper or cardboard. Or, divide a piece of paper into thirds with a pencil. Let the child paint the red and blue stripes.

   The French flag is called the tricolor. In 1789, King Louis XVI first used its three colors to represent France. Red, white, and blue have since come to represent liberty, equality, and fraternity – the ideals of the French Revolution.

2. Color or make a map of France.

3. Plant a tree.

4. Make Casse-croute.

2 slices of bread	1 slice Gruyere cheese
1 slice of ham	butter

   Directions: Prepare as you would grilled cheese.

5. Make Quiche Lorraine.

6 slices crisp bacon, crumbled	8-inch piecrust
3 eggs	1 3/4 cup cream or half & half
1/4 cup grated Swiss cheese	1/2 teaspoon salt
dash of pepper	pinch of nutmeg
1 tablespoon butter	

   Directions: Cover bottom of pie crust with bacon. Beat together eggs, cream, Swiss cheese, and seasonings. Pour over bacon. Dot with butter and bake at 375° for 30 minutes.

6. Make chocolate truffles.

6 oz. unsweetened chocolate	1 cup butter
4 teaspoons heavy cream	1 cup confectioner's sugar
4 tablespoons nut, finely chopped	cocoa powder

Directions: In a saucepan over low heat, melt the chocolate and butter. Stir in cream and gradually add sugar and nuts. Stir constantly until mixture is smooth. Transfer to a bowl; cover and refrigerate several hours. Form into small balls. Roll in cocoa. Store in refrigerator.

7.   Make perfumed beads.

6 handfuls dried flowers	medium saucepan
water	blender or food processor
colander with tiny holes	cookie sheet
needle	string

Directions: Place dried flowers in saucepan. Add just enough water to cover. Simmer on low for one hour (do not boil). Add more water, if necessary. Pour flowers and water into blender. Blend for 1-2 minutes. Pour mixture into colander. Squeeze excess water out with your hands. Roll mixture into small beads. Place on cookie sheet. Place the beads in a warm place to dry for 3-4 days. Carefully push the needle through dried beads. String the beads.

Internet Resources

❑ www.louvre.fr/louvrea.htm

❑ www.franceway.com   -   Click on "Paris" to see index of photographs

❑ http://encarta.msn.com/find/MediaList.asp?pg=6&mod=2&ti=761568934

Bible

❑ *The Giraffe That Walked to Paris*
  • 2 Corinthians 5:18
  • Matthew 5:23-24
  • Proverbs 25:11
  • Proverbs 16:21
  • James 1:4
  • Psalm 40:1
  • Colossians 1:11
  • Proverbs 12:10

❑ *Bon Appetit, Bertie!*
  • I Corinthians 10:31

❑ *Mirette on the High Wire*
  • Jeremiah 29:11

❑ *Marie in Fourth Position*
  • Colosians 1:9-11
  • Romans 5:3
  • James 1:3
  • I Thessalonians 5:14
  • Ecclesiastes 7:8

  • Romans 2:7

❑ *Madeline*
- I Corinthians 14:40
- Philippians 4:8
- Proverbs 1: 10
- Colossians 3:12
- Acts 9:36
- I Peter 3:10-12

❑ *A New Coat for Anna*
- Mark 12:31-44
- Proverbs 23:5
- Matthew 6:19
- Exodus 36:7-8
- Matthew 6:8
- Philippians 4:19

❑ *Linnea in Monet's Garden*
- Proverbs 12:27
- Colosians 3:23
- Matthew 25:1-30

❑ *The Glorious Flight*
- Psalm 128
- Psalm 127
- 2 Chronicles 15:7

❑ Trees
- 2 Chronicles 2:8 & 9
- I Kings 10:11 & 12
- I Kings 6:23, 31, & 33
- Joel 1:12
- Isaiah 44:13-20
- Isaiah 60:13
- Ezekiel 17:22-24
- Ezekiel 27:4-6
- Ezekiel 31:1-9
- Isaiah 2:12-18
- Luke 13:6-9
- Judges 9:8-15

- 2 Samuel 6:5
- Psalm 1
- Psalm 104:16-17
- I Chronicles 14:14-15
- Nehemiah 8:15-16
- Zechariah 1:8-11
- Genesis 8:10-12
- James 3:10-12
- John 12:12-13
- Isaiah 41:19
- Luke 19:1-10
- Genesis 2:8-9

Flag of France

Crossword Puzzle

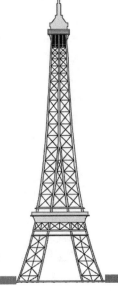

Across

   1 Capital of France.
   5 Famous landmark in 1 across.
   6 First name of girl burned at the stake.
   7 Scientist that worked with radium.

Down

   1 France is known for this product.
   2 Famous painter.
   3 Scientist who studied microbes.
   4 Language spoken in France.

Italy

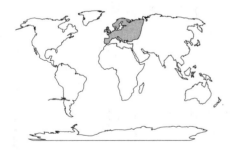

Population_____

Capital City_____

Religion_____

Type of Government_____

Currency_____

Language_____

What are the people called?_____

Italy

Italy acquired its name from the ancient Romans. The Romans called it "Italia", meaning "land of oxen" or "grazing land".

Italians take great pride in their cooking. They eat their main meal between noon and 3:00 p.m. Antipasti, or appetizers, are served first. Then a pasta or soup is served with crusty bread (no butter). Next is usually a meat dish with a green salad. Fruit, cheese, and sometimes, dessert follow the meal. Traditionally, wine is served with every meal except breakfast. Italians love ice cream. In fact, they invented it! Gelato is sold on city streets from specially made motorcycle carts. Ancient Romans made iced wines and fruit juices and learned how to preserve ice in the summer. Ices have been popular in Italy ever since.

On the day before Easter in Florence, Italy, a famous display takes place. It is called Scoppio del Carro or "Explosion of the Car". After two solemn days of mourning the death of Jesus, a rocket shaped like a dove is lit at the church alter. It shoots along a wire and through the doors of the church to the car. The crowds cheer wildly as the car explodes, signifying Jesus rising from the dead.

People/History

- ❑ Michelangelo
  - *Getting to Know the World's Greatest Artists: Michelangelo* by Mike Venezia
  - *Michael the Angel* by Laura Fischetto
  - *Michelangelo's Surprise* by Tony Parillo

- ❑ Leonardo Da Vinci
  - *Getting to Know the World's Greatest Artists: Da Vinci* by Mike Venezia
  - Pages 9, 45, & 66 – *Usborne Book of Famous Lives*
  - *Leonardo Da Vinci* by Diane Stanley

- ❑ Galileo
  - *Starry Messenger* by Peter Sis
  - Pages 8,66, & 67 – *Usborne Book of Famous Lives*

- ❑ Giotto
  - *Getting to Know the World's Greatest Artists: Giotto* by Mike Venezia
  - *A Boy Named Giotto* by Paolo Guarnieri

General Reference Books

- ❑ *Passport to Italy* by Cinzia Mariella
- ❑ *A Family in Italy* by Penny Hubley
- ❑ *Italy* by Kristin Thoennes
- ❑ *Look What Came From Italy* by Miles Harvey
- ❑ *Next Stop Italy* by Clare Boast
- ❑ *Italy* by Mary Berendes

Literature

- ❏ *Papa Piccolo* by Carol Talley
- ❏ Buried Treasure – *Stories From Around the World* by Usborne
- ❏ *St. Francis & the Friendly Beasts*
- ❏ *Antonio's Apprenticeship* by Taylor Morrison
- ❏ *Clown of God* by Tomie de Paola

Science

- ❏ If studying Galileo, do a unit on astronomy.
  - Lesson 2 – *Considering God's Creation* - The Universe
  - Lesson 3 – *Considering God's Creation* - Our Solar System
  - *Looking at the Planets* by Melvin Berger
  - *Find the Constellations* by H.A. Rey
  - *The Sun* by Kate Petty
  - Journeys to the Edge of Creation – videos by Moody Institute of Science

Internet Resources

- http://seds.1pl.arizona.edu/nineplanets/nineplanets/
- www.windows.ucar.edu

Vocabulary

canal	opera	pasta
gondola	colosseum	sculpture
architecture	mosaic	fresco

- ❏ If doing a unit on astronomy:

orbit	planet	astronaut
moon	star	telescope
asteroid	meteorite	galaxy
comet	astronomy	

Music/Art/Projects

1. Color or make the flag of Italy. See France for ideas (use red, white, and green).

   Adopted in 1870, the Italian flag was first used in 1796 by Italians who supported Napoleon of France during a war against Austria. Napoleon designed the flag to look like the French flag, but substituted his favorite color green for the blue of the French flag.

2. Color or make a map of Italy.

3. Learn 1-10 in Roman numerals.

4. Paint

   If studying Michelangelo, tape a large piece of paper under a low table and have the children paint while lying on their back. (Recommend doing this activity outside!)

5. Listen to The Story of Verdi – Music Masters Series.

6. Make Gelato (Italian ice).

   | 4 cups water | 2 1/2 cups sugar |

   2 cups orange, lemon, or cranberry juice

   Directions: In a medium saucepan, bring water and sugar to boil. Stir until sugar is dissolved. Boil for five minutes, stirring often. Cool completely. Stir in juice. Pour into freezer container and freeze for 4-5 hours. Stir occasionally.

7. Make Italian food.
   - *Cooking the Italian Way* by Alphonse Bisignano
   - *Italian Food and Drink* by Edwina Biucchi

Internet Resources
- ☐ www.mi.cnr.it/WOI/images/pictures.html
- ☐ www.wandering.com
- ☐ www.compart-multimedia.com/virtuale/us/home.htm
- ☐ http://encarta.msn.com/find/MediaList.asp?pg=6&mod=2&ti=761555207

Bible
- ☐ *Papa Piccolo*
  - Psalm 68:5
  - James 1:27
- ☐ *Antonio's Apprenticeship*
  - Romans 12:11
  - Ecclesiastes 9:10
  - Colossians 3:22
- ☐ *Clown of God*
  - Proverbs 19:17
  - Psalm 103:2

- ☐ If studying astronomy:
  - Genesis 1:14-19
  - Job 9:9
  - Job 38:31-33
  - Amos 5:8
  - Jeremiah 31:35
  - Psalm 147:4
  - 1 Corinthians 15:41
  - Matthew 2:2, 9, 10
  - Joshua 10:12-14
  - Ecclesiastes 1:5
  - 2 Samuel 23:4
  - Psalm 19:6

Flag of Italy

FIND THE TWINS

Which two are exactly alike?

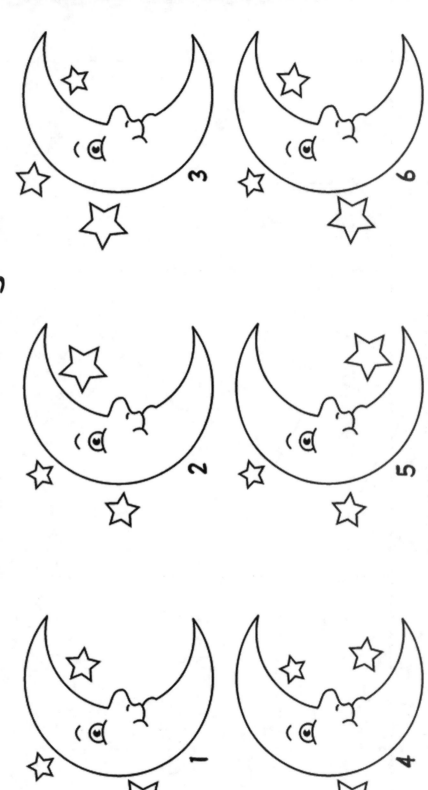

MAZE CRAZE

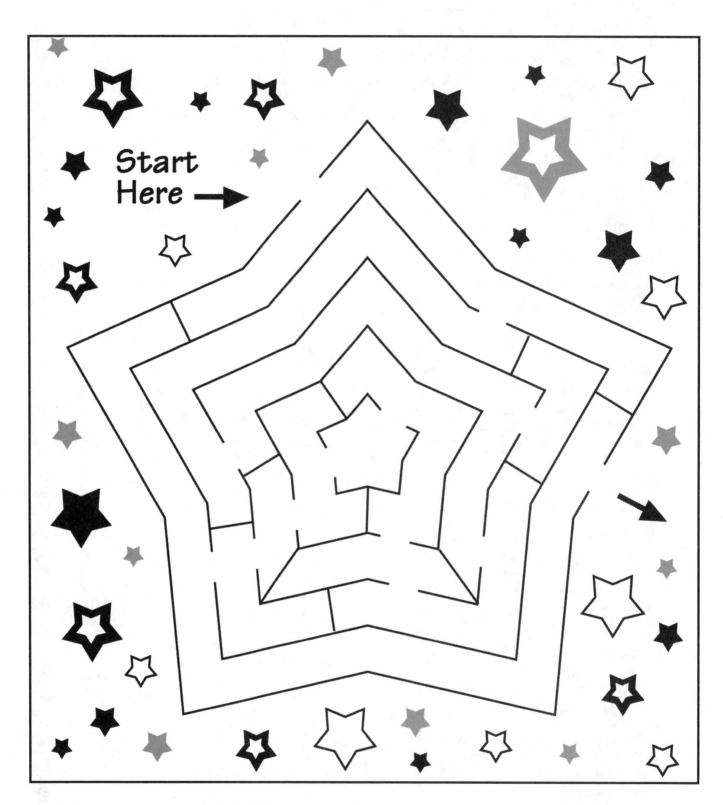

Start
Here →

Word Search

```
p l u t o n q a s e k u c k r v z x o k
m a r s v f a o d x f d s z n z d l r v
w o h z a s t e r o i d f v g h a a b t
f m m w k h y m w p a m v y o p j e i n
s i y m e r c u r y e k j o p l j d t l
r l b o h w b s w e r b l f k a y d j u
w k s x m a r x v z m o o n o n j g p s
s y i q e t u r k j j c w u x e q p o o
h w f c t f v u g z c m h e a t j g x l
r a m q e d n e p t u n e k e p s z e a
s y z p o y t g p k j k f p o p s v n r
f c k x r y o i e y t e s c x w a q f s
v i h d i p x e a r t h a o t j t d v y
a n d f t i d r k d x a n m e o u s z s
c e q w e q a s t a r l e e w j r b b t
q z y l t a d b n g k r f t b y n q p e
i u r a n u s v p c l d w h z t m t o m
k x y y y z o i d u f p k a e m d o b u
o x h b n g n c e m j u p i t e r f a v
v e n u s e y t x g z w x i v y p e y q
```

asteroid
comet
Earth
Jupiter
Mars
Mercury
meteorite
Milky Way
moon

Neptune
orbit
planet
Pluto
Saturn
solar system
star
Uranus
Venus

Crossword Puzzle

Across

1 The second planet from the sun.
3 The red planet.
4 The seventh planet from the sun.
7 Our solar system.
8 The largest planet.
9 This planet has rings.
10 The planet closest to the sun.
12 Any of the thousands of small planets that orbit the sun, mostly between Mars and Jupiter.
13 A heavenly body that orbits around a star, such as the sun.
14 A mass of ice, frozen gasses, and dust particles that travels around the sun in a long, slow path.

Down

2 The eighth planet from the sun.
3 A heavenly body that revolves around a planet.
5 The sun together with the planets and other heavenly bodies that it orbits.
6 A heavenly body that shines by its own light.
11 The third planet from the sun.
13 The smallest planet.

Germany

Hundreds of years ago, Germans created many famous dishes to prevent food from spoiling. Sauerkraut was developed to preserve cabbage. Sausages were made to preserve meats. Sauerbraten was invented by soaking meat in spices and vinegar. There are over 1,500 types of German sausage.

Each autumn, Oktoberfest is celebrated in Munich with parades, fancy costumes, and refreshments. The festival originally began in the 1800's to celebrate a royal wedding.

Germans have developed some of the most famous inventions: the printing press, pocket watches, bicycles, x-ray machines, electric trains, and the Fahrenheit thermometer scale.

There are many castles in Germany, especially along the Rhine River. Neuschwanstein castle in Southern Germany inspired the castle in Walt Disney's "Sleeping Beauty".

People/History

- ❑ Ludwig Van Beethoven
    - *Ludwig Van Beethoven: Musical Pioneer* by Carol Greene
    - *Beethoven: Getting to Know the World's Greatest Composers* by Mike Venezia
    - Beethoven Lives Upstairs – Classical Kids Audio

- ❑ Johannes Gutenberg
    - Page 26 – *Usborne Book of Famous Lives*
    - *Gutenberg* by Leonard Everett Fisher

- ❑ Albert Einstein
    - Page 88 – *Usborne Book of Famous Lives*
    - *Young Albert Einstein* by Laurence Santrey
    - *Albert Einstein* by Ibi Lepscky

- ❑ Johann Sebastian Bach
    - Mr. Bach Comes to Call – Classical Kids Audio
    - *Getting to Know the World's Greatest Composers: Bach* by Mike Venezia
    - *Sebastian: A Book About Bach* by Jeanette Winter

- ❑ Brahms
    - *Getting to Know the World's Greatest Composers: Brahms* by Mike Venezia
    - The Story of Brahms – Music Masters Series

- ❑ Handel
    - *Handel and the Famous Sword Swallower of Halle* by Bryna Stevens
    - *The Duke's Command* by Phyllis Berk
    - Hallelujah Handel – Classical Kids Audio

Germany

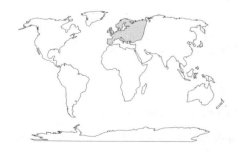

Population_____

Capital City_____

Religion_____

Type of Government_____

Currency_____

Language_____

What are the people called?_____

❑ Castles
  • *Castles* by Gillian Osband & Robert Andrew
  • *Castles* by Gallimard Jeunesse
  • *Castle Life* by Struan Reid
  • *Incredible Castles and Knights* by Christopher Maynard
  • *How Castles Were Built* by Peter Hicks

General Reference Books

❑ *Usborne's First Book of Germany*
❑ *Count Your Way Through Germany* by Jim Haskins
❑ *Picture A Country: Germany* by Henry Pluckrose
❑ *Next Stop Germany* by Clare Boast
❑ *Germany* by Mary Berendes
❑ *A Family in West Germany* by Ann Adler

Literature

❑ Musicians of Bremen – *Stories From Around the World* by Usborne
❑ *Goldilocks and the Three Bears* retold by Armand Eisen
❑ *The Pied Piper*
❑ *Heidi* by Johannes Spyri
❑ *The Duchess Bakes a Cake* by Virginia Kahl
❑ *Bach's Big Adventure* by Sallie Ketcham

Science

❑ Lesson 10a - *Considering God's Creation* - Plants

Vocabulary

German words used in English:

delicatessen	pretzel	kindergarten
hamburger	sauerkrat	sausage

Music/Art/Projects

1. Color or make the flag of Germany. See France for ideas (use red, black, and yellow).
    The colors of the German flag are associated with the unification of Germany and the colors worn by German soldiers in the early 1800's.

2. Color or make a map of Germany.

3. Listen to music of the composers you have studied.

4. Make German food.

   Cooking the German Way by Helga Parnell

5. If you read *The Duchess Bakes a Cake*, bake a cake.

Internet Resources

❑ http://encarta.msn.com/find/MediaList.asp?pg=6&mod=2&ti=761576917

Bible

❑ The Bible has a lot to say about music and instruments. Studying the composers is a good time to study what God has to say about music.

- 2 Samuel 6:5
- 1 Chronicles 6:32
- 1 Chronicles 13:8
- 1 Chronicles 15:11-16:43
- 1 Chronicles 25:6
- 2 Chronicles 5:13
- 2 Chronicles 7:6
- 1 Samuel 10:6
- Nehemiah 12:27

- Psalm 98:4-6
- Psalm 147:7
- Psalm 33:1-3
- Exodus 15:20 & 21
- Psalm 96:1
- Psalm 150
- Psalm 149:1-5
- Isaiah 38:20
- Habakkuk 3:19

❑ *The Duchess Bakes a Cake*
- Galatians 6:7

Flag of Germany

85

FIND THE TWINS

Which two are exactly alike?

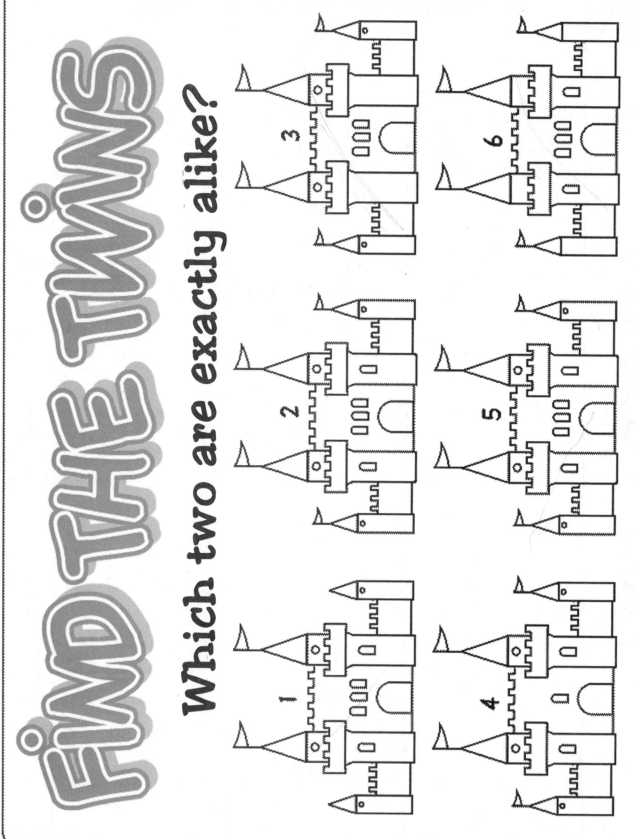

1

2

3

4

5

6

MAZE CRAZE

**Start
Here**
Find the
black note.

Holland

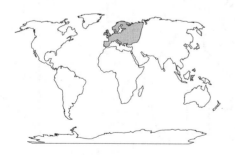

Population_____

Capital City_____

Religion_____

Type of Government_____

Currency_____

Language_____

What are the people called?_____

Holland

Holland is about the size of the state of Maryland. Amsterdam is the official capital, but the official residence of the Queen and the headquarters for the national government is The Hague (known as the Peace City). Holland is also known as The Netherlands.

More than a fourth of the land in Holland is below sea level. Huge earthen dikes, or dams, keep the water from overflowing onto the land. Some dikes have a narrow foot or bicycle path along the top of it. Others are so wide that highways or railroads have been built on top of them. Windmills can be found throughout Holland. The Dutch use the windmills as a source of energy for grinding grain and pumping water. Today, electric motors have replaced most windmills.

Bicycling is a favorite means of travel in Holland. At each side of the main roads there are smooth bicycle paths. Celebrating birthdays is important to the Dutch. Special activities are planned all day long to honor a family member's birthday. Milk, cheese, and flowers are among Holland's most famous products. Several thousand different kinds of tulips and other flowers are grown in the country.

People/History

❑ Rembrandt

- *Rembrandt: Getting to Know the World's Greatest Artists* by Mike Venezia

❑ Vincent Van Gogh

- *Art for Young People: Vincent Van Gogh* by Peter Harrison
- *Van Gogh* by Bruce Bernard (Eyewitness Books by Dorling Kindersley) – Note: There are a few nude paintings in this book.
- *Van Gogh* by Josephine Cutts & James Smith – Like the book *Monet* by Vanessa Potts, this book has a photograph of Van Gogh's work on one page and a short discussion or background on the work on the opposite page. Note: Two of the works shown portray a woman's nude backside.

General Reference Books

❑ *Dutch Low Country – Countries and Their Children*
❑ *The Netherlands* (A New True Book) by Karen Jacobsen
❑ *Take a Trip to Holland* by Chris Fairclough

Literature

❑ *Boy Who Held Back the Sea* by Lenny Hort
❑ Brave Hendrick – *Stories From Around the World* by Usborne
❑ *Hans Brinker or the Silver Skates* by Mary Mapes Dodge
❑ *The Hole in the Dike* by Norma B. Green
❑ *The Cow Who Fell in the Canal* by Phyllis Krasilovsky
❑ *The Little Riders* by Margaretha Shemin

Science

☐ Flowers

  • Lesson 10b – *Considering God's Creation* - Flowers
  • *Flowers* by Gail Saunders-Smith
  • *Flowers* by Gallimard Jeunesse
  • *Flowers* by David Burnie

Vocabulary

| dike | windmill | cheese |
| cleanliness | polder | barge |

☐ If studying flowers:

stem	stamen	pollen
petal	root	
pistil	leaf	

Music/Art/Projects

1. Color or make the flag of Holland.
    The flag of Holland has three horizontal stripes of red, white, and blue. Before 1630, an orange stripe was at the top instead of red to signify the House of Orange.

2. Color or make a map of Holland.

3. Taste Dutch cheeses (Edam, Gouda, etc.)

4. Some cities in Holland have miniature model villages. Use small cartons and boxes to make a miniature village.

5. Go bowling.

6. Make paper windmills.

7. Make cheese.
    Making Cheese, Butter, & Yogurt by Phyllis Hobson is the best resource we have found for this project. It tells you how to make the cheese form and press from materials found around the house. *Cheese Making Made Easy* by Ricki Carroll & Robert Carroll and *Making Great Cheese* by Barbara Ciletti are also good.

8. Make Dutch Boiled Dinner

2 lb. beef brisket	1 1/2 cups water
1 teaspoon salt	4 medium potatoes, diced
4 medium carrots, sliced	3 medium onions, chopped
1 1/2 teaspoons salt	1/4 teaspoon pepper
snipped parsley	prepared mustard or horseradish

Heat beef, water, and 1 teaspoon salt to boiling in Dutch oven; reduce heat. Cover and simmer 1 1/2 hours. Add potatoes, carrots, onion, 1 1/2 teaspoons salt, pepper. Cover and simmer until beef and vegetables are tender, about 45 minutes. Drain meat and vegetables, reserving broth. Mash vegetables; mound on heated platter. Cut beef across grain into thin slices; arrange around vegetables. Garnish with parsley. Serve with reserved broth and mustard or horseradish.

9. Make Chocolate Bread

bread	butter	chocolate chips

Directions: Spread bread with butter; sprinkle with chocolate chips. Heat in oven or microwave, if desired, or serve cold.

Internet Resources

❑ http://encarta.msn.com/find/MediaList.asp?pg=6&mod=2&ti=761572410

❑ www.wikipedia.com/wiki/The_Dutch_monarchy

❑ www.odci.gov/cia/publications/factbook/geos/nl.html

Bible

❑ Cheese
  • Samuel 17:29
  • Job 10:10
❑ If studying flowers:
  • Song of Solomon 6:2
  • Song of Solomon 2:1
  • Isaiah 35:1 & 2
  • Matthew 6:28 & 29
❑ *Boy Who Held Back the Sea*, *Brave Hendrick*, & *The Hole in the Dike*
  • Psalm 46:1-3
  • Psalm 91:5a
  • Psalm 118:6a
❑ *The Cow Who Fell in the Canal*
  • Psalm 144:15
  • Proverbs 3:13
  • Proverbs 14:21
  • Proverbs 16:20
  • Proverbs 28:1

HOLLAND

Flag of Holland

MAZE CRAZE

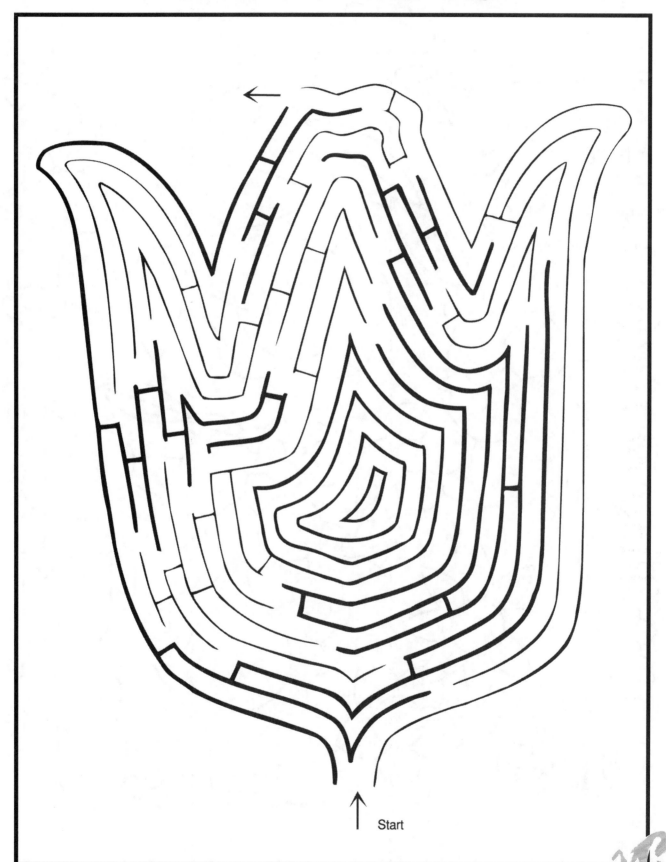

Start

MAZE CRAZE

Find the bee.

Start Here
↓

94

FIND THE TWINS

Which two are exactly alike?

1

2

3

4

5

6

Spain

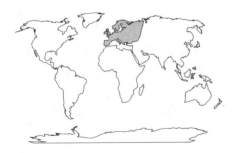

Population_____

Capital City_____

Religion_____

Type of Government_____

Currency_____

Language_____

What are the people called?_____

Spain

Spain dominates the Iberian Peninsula. Portugal occupies the southwestern part of the peninsula, facing the Atlantic Ocean. Andorra is a little state in the Pyrenees, while Gibraltar is a British colony. The Spanish capital is Madrid, which is located right in the middle of the country, on the hot Meseta. Spain's second city is Barcelona, a lively cultural and commercial center on the Mediterranean coast.

Over three-fourths of the Spanish population live in cities and towns. Many of these include historical centers with impressive castles and cathedrals as well as industrial and residential suburbs. The town of Bunol in the eastern province of Valencia, Spain, holds an unusual festival each August. During the Tomatina, people pelt each other with tomatoes!

Spain is one of Europe's major car manufacturers. Vehicles are the country's most important export. Spain enjoys a warm climate. This means that crops such as lemons, oranges, olives, and melons can be grown. Cork is obtained from a type of oak tree. Large fishing fleets catch sardines, tuna, anchovies, and cod.

People/History
- ❑ Queen Isabella
  - • Page 197 – *Usborne Book of Famous Lives*
  - • *Queen Isabella I* by Corinn Codye
  - • *Isabella of Castile* by Joann J. Burch

General Reference Books
- ❑ *Discovering Spain* by Philippa Leahy
- ❑ *Spain* by Catherine Chambers
- ❑ *Spain* by Mary Berendes
- ❑ *Next Stop Spain* by Clare Boast
- ❑ *Spain* by Kate A. Furlong
- ❑ *Spanish Food and Drink* by Maria Eugenia D. Pellicer

Literature
- ❑ *Story of Ferdinand* by Munro Leaf

Science
- ❑ Bulls: If your library, like ours, does not have books specifically on bulls, look for books on cows or cattle.
  - • *Cows* by Mary Ann McDonald
  - • http://encarta.msn.com/find/MediaList.asp?pg=6&mod=2&ti=761573997

- ❑ Cork
  - • www.dicknsons.com/corkhist.htm
  - • www.granorte.pt/cork_oak.htm

Notebook Suggestions:

1. What is cork?

2. Explain what cork is used for.

3. How long does a cork tree live?

4. How old is the tree before it produces quality cork?

Vocabulary

bull	bullring	cork
dance	olive	seaport

Music/Art/Projects

1. Color or make the flag of Spain.

   Red and yellow are the colors of the arms of both Castille and Aragon. The first red and yellow flag of Spain was adopted in the 18th century for use at sea. The present layout was adopted in 1975.

2. Color or make a map of Spain.

3. Make Baked Fish, Spanish Style

1 1/2 lbs. fish steaks or fillets	1 1/2 teaspoons salt
1/4 teaspoon paprika	1/4 teaspoon pepper
1 green pepper, cut into rings	1 sliced tomato
1 small onion, sliced	2 tablespoons lemon juice
2 tablespoons olive oil	1 clove garlic, minced
Lemon wedges	

   If fish pieces are large, cut into serving sizes. Arrange fish in ungreased square baking dish; sprinkle with salt, paprika, and pepper. Top with green pepper rings and tomato and onion slices. Mix lemon juice, oil, and garlic; pour over fish. Cover and cook in 375° oven 15 minutes. Uncover and cook until fish flakes easily with fork, 10 to 15 minutes longer. Garnish with lemon wedges.

Internet Resources

☐ www.red2000.com/spain/index.html

☐ http://encarta.msn.com/find/MediaList.asp?pg=6&mod=2&ti=761575057

Bible

☐ *Story of Ferdinand*
- Mark 1:35

☐ Bulls
- Exodus 32
- Psalm 22:12
- Hosea 10:11
- Jeremiah 50:11
- Genesis 41:1-7

Flag of Spain

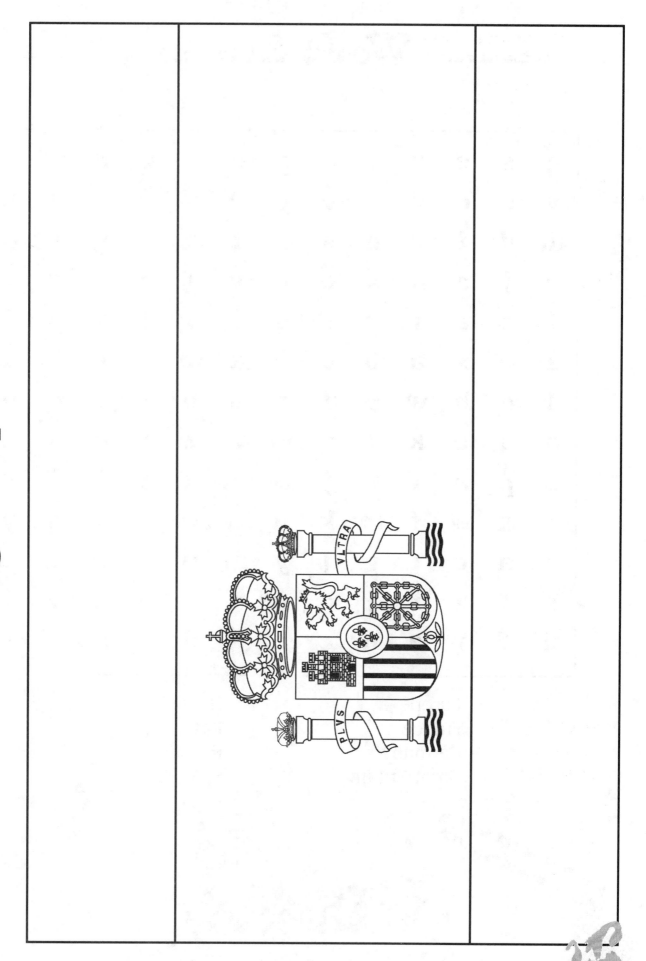

Word Search

```
q s e u r o p e f g n m g
v u o k v c g d y r s p h
m d i r u s s i a e g r o
v j a k x u p v j a b r l
i t a l y b g i j t z p l
z o x a b d e a m b g f a
l q b w p i r u h r s o n
b q e k l a m w z i o a d
s p a i n j a i x t j l r
l x w f m k n n e a i q y
b a c t g j y m o i d r c
s f z r j o g t c n u v i
a f r a n c e x d e m q r
```

Europe   Holland

France   Italy

Germany   Russia

Great Britain   Spain

Spain and Portugal

What are people from Spain called? _____

What language do they speak? _____

What are people from Portugal called? _____

What language do they speak? _____

Europe Review Map

See how many countries you can identify. Write their names on the map.

© 2003 Geography Matters

ad maiorem Dei gloriam!

Christmas Around the World

Christmas is the most celebrated holiday in the world. Plan on taking the month of December to relax and enjoy learning about the different and wonderful ways Christmas is celebrated around the world. Every country's celebrations vary according to the climate, beliefs, traditions, and folklore of that country. Some countries enjoy some of the same practices and symbols.

A fun activity to consider is setting up a small Christmas tree in your school area. When you study a different country add an ornament to remind your children of the way that country celebrates Christmas.

A very brief description of how 10 different countries celebrate the holiday season follows. Choose the ones you would like to learn more about then check out some of the reference books suggested to find recipes, activities, and music from that individual country.

Japan

Most Japanese are not Christians; therefore, the majority of the people do not celebrate the religious aspects of the holiday. The Japanese Christmas greeting is "Meri Kurusumasu". Japan has adopted many western Christmas traditions such as exchanging gifts, caroling, and decorating store windows and homes with holly, and bells. The Japanese use lanterns, fans, flowers, and dolls to decorate their trees. Hoteiosho is the Japanese Santa Claus. He walks around and observes children with "eyes in the back of his head". If the children are good he gives them a toy from the bag he carries.

❑ Activity: Make small Japanese fans to decorate your international tree.

China

The Chinese New Year is the biggest celebration in China. During this time the people enjoy gift-giving and fireworks for an entire week. Chains, flowers, and paper lanterns are popular Chinese decorations. Special lanterns shaped like pagodas sometimes show the Holy Family inside. A special gift-bearer, Lan Khoong-Khoong or Nice Old Father, fills the children's stockings with small gifts. Another name for the gift-bearer is Dun Che Lao Ren, the Christmas Old Man.

❑ Activity: Make a paper chain for your international tree.

Russia

The Christmas greeting in Russia is "Hristos Razdajetsja". Dyed Maroz, Grandfather Frost, is the Russian Santa Claus. He dresses in a red suit and has a white beard, but delivers gifts on New Year's Day. One traditional gift is a Matryoshka doll. The outer doll is opened to reveal smaller dolls nested inside. Another gift-bearer famous in Russia is Babouschka. The legend says that she was visited by the Three Kings, but she was too busy to direct them to the Christ child. Because of her error she is doomed to wander forever and deliver gifts to good children. She pay's her visit on Epiphany, January 6. Russian families eat Christmas Eve supper together and decorate their tree with candy, oranges, apples, dolls, fabric, and foil ornaments.

❑ Activity: Add some apples or oranges to your international tree.

Great Britain

"Happy Christmas" is the English Christmas greeting. Many of the most popular Christmas customs originated in Great Britain. The first Christmas cards were sent in England. Tradition says that boarding school children would send them to their parents. Caroling is another British custom. It is like American caroling, except groups of people sip wassail, a hot punch-like drink, while walking up and down the street. Wassail means "be in good health". This event takes place all throughout the 12 days of Christmas (December 25 – January 6). Decorations include holly, ivy and mistletoe. The British also have Christmas "crackers", small circular shapes filled with small prizes that make a cracking sound when opened. Christmas dinner often includes 12 or more courses. December 26 is Boxing Day. This is when people rewarded good servants with gifts, today community workers often receive this special appreciation.

❑ Activity: Add holly, ivy, or mistletoe to your international tree.

France

In France the season begins on December 5, St. Nicholas Eve. This is one of several days that children receive gifts. The children leave their shoes by the fireplace in hope of receiving special treats. It is believed that France is the first country to begin leaving gifts in the name of St. Nicholas. This custom is popular in many other countries today. Christmas Eve is when parents leave toys, fruit and candy for the children to find the next day, usually these are left on the branches of the Christmas tree. New Year's Day is when friends and family members give gifts. The French also made the manger scene popular. Christmas trees are decorated with stars of many different colors and the crèche (manger scene) is the center of the decorations. A few days before Christmas the family carefully assembles the crèche and decorates it with evergreens and candles. Then they celebrate the birth of Christ by singing carols and rejoicing. Food is very important at the French celebration. Foods like Buche de Noel, and Galette des Rois are made this time of year.

❑ Activity: Make paper stars for the international tree.

Italy

Christmas in Italy is a solemn, yet festive, occasion. The first manger scene originated in Italy and was made by St. Francis of Assisi to encourage others to worship Jesus. Italy is credited with the first true Christmas carols. The Italian Christmas begins on the first Sunday of Advent. The nine days before Christmas include bagpipers in the streets, fireworks, bonfires, carols, and lots of lights. The manger scene or presepio is set up in homes without baby Jesus, then on Christmas Eve the figure is passed around and put in the manger with songs and prayers. At 10:00 Christmas Eve mass begins. January 6 is the traditional gift-giving day; however, some families give gifts on Christmas Day as well. These gifts come from Gesu Bambino or Baby Jesus. Christmas Day is a day of church, family and feasting. The family eats pasta dishes and turkey. On January 6, children receive gifts from La Befana. Legend describes her as a tiny old woman who is dressed in black and rides on a broomstick. Tradition says the Befana was visited by the Three Kings in search of Jesus. They asked her to come along, but she was too busy working. She went to go with them later, and they were already gone. She still searches for them today. She leaves gifts for the good children and ashes or coal for the naughty ones.

❑ Activity: Put a manger scene under the international tree.

Germany

Merry Christmas in German is "Frohliche Weihnachten". One of our most important traditions, the Christmas tree, originated in Germany. In the sixteenth century, Martin Luther was the first person to bring an evergreen indoors. While he was outside on Christmas Eve he was so moved by the tree against the starlight that he cut one down and brought it home. He put lighted candles on it to symbolize the stars over Bethlehem. Families would originally light trees only on Christmas Eve. They decorated them with apples, cookies, candies and candles. The tradition of the Christmas tree spread to France and England, they are credited with adding the angel on top. There are many gift-bearers in Germany depending in which region you are. Each one is a helper to Kirst Kindl or Christkindl, the Christ Child. These gifts are brought by a young child dressed in white wearing a crown of candles. In other parts of Germany Kris Kringle is the giftgiver. He may arrive by mule or white horse. The children leave goodies for him to eat and he in turn leaves gifts for the good children. If the children have been bad then Hans Trapp will leave them switches.

Families in Germany use the advent wreath and light one candle each Sunday. The main celebration is on Christmas Eve, this is when the Gemans decorate the tree, go to church, eat, sing, and give gifts.

❑ Activity: Add dried apples or apple shaped ornaments to the international tree.

Spain

"Felices Pascuas" is Merry Christmas from Spain. Religious celebrations are predominant in mostly Catholic Spain. In Spain, Christmas is celebrated from December 24 - January 6. Public areas are decorated with life-size nativity scenes. Plays depicting the shepherds adoring Jesus are popular events. Spain uses many lights in its decorations. Christmas Eve is the "Good Night" and people fast all day, not eating until after midnight. Bells chime loudly at midnight calling people to midnight mass. After mass the feasting begins. They have paella, (a rice and seafood dish), fruits, candy and chirimoyas (apple custard). Christmas Day is spent with friends and family. The Three Kings deliver gifts on the eve of January 5. The children leave out their shoes and find them filled with toys and treats the next morning. On January 6 there is a parade with kings and animals up and down the streets.

❑ Activity: Add bells to your international tree.

Mexico

"Feliz Navidad" is the Christmas greeting. Like Spain, Mexico is mostly Catholic. The Processions of Las Posadas are very important in Mexico. These processions symbolize the travels of Mary and Joseph to Bethlehem. People divide into two groups, the innkeepers and the travelers. Stopping places are chosen ahead of time. At each stop the travelers are denied entry until the last stop, where the group is invited in for eating and celebrating. These processions occur December 16-24. The final celebration includes fried sugar tortillas, hot chocolate, and a piñata for the children to break open. After nine evenings of celebrating, Christmas Day is a quiet family time with feasting and reflection. On January 6 the children receive their gifts from the Three Kings. They leave their shoes out to be filled with gifts and treats. The poinsettia is native to Mexico where it grows wild in damp areas. In 1829 the U.S. ambassador to Mexico brought the plant home to the US.

❑ Activity: Play "innkeepers and travelers" using different rooms of your home.

                    Add silk poinsettias to the international tree or place a poinsettia plant beside tree.

Nigeria

Almost half of Nigerians are Christians; therefore, many Christian customs are practiced in Nigeria. Large church pageants are performed and used to spread the gospel to non-believers. Instead of gifts, Nigerians often give food on Christmas Day. They make extra of their favorite dishes and send portions to their friends and neighbors. Christmas afternoon, children go from house to house singing carols anticipating candy or cookies in return. Music is important in Nigeria and the drum and other instruments are used extensively in Christmas celebrations.

❑ Activity: Have children create their own Christmas pageant using lots of music.

People/History

☐ St. Nicholas

For centuries St. Nicholas has been associated with Christmas and gifts. His name, originally from the Latin, Sanctus Nicolaus, has had various forms, including the German, Sankt Nikolaus, Dutch Sinter Klaas, finally becoming our modern "Santa Claus". Although he is regarded as a myth, there actually was a real St. Nicholas, an early Christian who lived during the fourth century.

Nicholas, the only child of wealthy Christian parents, was born at the end of the third century at Patara, a port in the province of Lycia in Asia Minor. From early childhood his mother taught him the Scriptures. When both parents died during an epidemic, they left the young boy in possession of all their wealth.

Young Nicholas dedicated his life to God's service and moved to Myra, the chief city of his province. One of Nicholas's best characteristics was his unsurpassed generosity. In his youth he had learned, by going around among the people, how many were oppressed by poverty. As a result, he often went out in disguise and distributed presents, especially to children. Stories of Nicholas's kindness and liberality soon spread. As a result, when unexpected gifts were received, he was given credit as the donor.

General Reference Books

☐ *Holiday Cooking Around the World* published by Lerner Publications
☐ *Christmas Cooking Around the World* by Susan Purdy
☐ *A Christmas Companion* by Maria Robbins & Jim Charlton
☐ *Christmas in (name of country)* series published by Worldbook
☐ *The Whole Christmas Catalog for Kids* by Louise Betts Egan
☐ *Christmas Crafts and Customs Around the World* by Virginia Fowler
☐ *Celebrating Christmas Around the World* by Herbert Werneke
☐ *A Christmas Companion: Recipes, Traditions and Customs from Around the World* by James Charlton
☐ *Silent Night: Its story and song* by Margaret Hodges

Literature

☐ *Ellis Island Christmas*
☐ *The Bird's Christmas Carol* by Kate Douglas Wiggin
☐ *The Christmas Tree Ship* by Jeanette Winter
☐ *A Christmas Tree in the White House* by Gary Hines
☐ *Christmas Tree Memories* by Aliki
☐ *Tree of Cranes* by Allen Say
☐ *Jotham's Journey* by Arnold Ytreeide
☐ *Papa's Christmas Gift* by Cheryl Harness
☐ *An Amish Christmas* by Richard Ammon
☐ *A Candle for Christmas* by Jean Speare
☐ *A Northern Nativity: Christmas Dreams of a Prairie Boy* by William Kurulek
☐ *The Best Christmas Pageant Ever* by Barbara Robinson

Science

❑ Deer and reindeer

- *Reindeer* by Emery & Durga Bernhard
- *Reindeer* (A New True Book) by Emilie U. Lepthien
- *Deer, Moose, Elk, & Caribou* by Deborah Hodge
- *All About Deer* by Jim Arnosky
- *Never Grab a Deer by the Ear* by Colleen Stanley Bare
- *White-Tailed Deer* (A New True Book) by Joan Kalbacken
- *Little Caribou* by Sarah Fox-Davies
- http://encarta.msn.com/find/MediaList.asp?pg=6&mod=2&ti=761564152
- http://encarta.msn.com/find/MediaMax.asp?pg=3&ti=761571587&idx=461526333

Notebook Suggestions:

1. How much do caribou weigh?
2. Explain where caribou migrate in winter.
3. List what caribou eat during summer.
4. What are white-tailed deer recognized by?
5. Only bucks grow _____ .
6. List what deer eat.
7. Explain how a fawn's coloring is different from an adult deer.

Vocabulary

wassail	holly	manger
gift	decorate	tradition
worship	advent	shepherd
parade	pageant	magi

Music/Art/Projects

1. Make lots of Christmas crafts.
2. Learn Christmas carols and sing them for others.

Internet Resources

❑ General

- www.californiamall.com/holidaytraditions/home.htm
- www.soon.org.uk/country/christmas.htm
- www.epicurious.com/e_eating/e04_xmas/euro.html
- http://christmas-world.freeservers.com//index.html

- ❏ France
  - http://gofrance.about.com/library/weekly/aa121200.htm
  - www.geocities.com/Athens/Acropolis/1465/Christmas/france.html
  - http://christmas-world.freeservers.com/france.html

- ❏ Italy
  - http://christmas-world.freeservers.com/italy.html

- ❏ Spain
  - http://christmas-world.freeservers.com/spain.html

- ❏ Germany
  - www.geocities.com/Tokyo/Island/6639/xmas.htm
  - www.pastrywiz.com/cookies/index.html

- ❏ Mexico
  - www.mexconnect.com/index.html

Bible

- ❏ Luke 2:1-40
- ❏ Deer
  - 2 Samuel 2:18
  - Genesis 49:21
  - Psalm 18:30-33
  - 2 Samuel 22:33 & 34
  - Psalm 29:9
  - Psalm 42:1
  - Isaiah 35:6
  - Habakkuk 3:19

MAZE CRAZE

FINISH

START

110

The Poles

The North Pole lies in the area known as the Arctic Circle. This bitterly cold region includes the northern parts of Asia, Europe, and North America.

The South Pole is on the continent of Antarctica. While several countries have laid claim to Antarctica, the international community recognizes none of these claims. The continent's only inhabitants are scientists from around the world.

Map of Antarctica

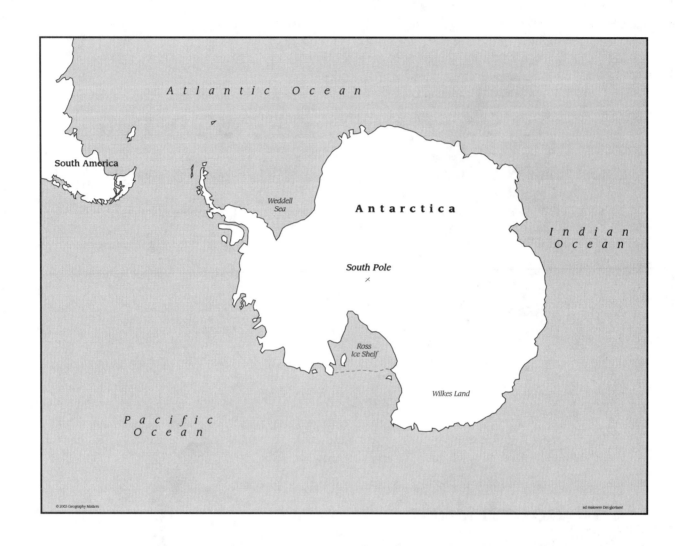

Atlantic Ocean

South America

Weddell
Sea

Antarctica

Indian
Ocean

South Pole

Ross
Ice Shelf

Wilkes Land

Pacific
Ocean

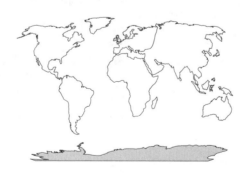

Antarctica

Antarctica is the fifth largest continent. It is twice the size of the United States. Its position at the South Pole and its elevation make it the coldest place on Earth. The coldest temperature ever recorded was 127 °F below zero! Antarctica receives only about five inches of snow per year, but because it is so cold, the snow never melts. Because the continent itself is barren, most life on and around Antarctica is supported by the sea.

People/History

❑ Roald Amundsen
  • Page 137 – *Usborne Book of Famous Lives*

❑ Ernest Shackleton
  • *Sea of Ice: Wreck of the Endurance* by Monica Kulling
  • Shackleton's Antarctic Adventure - video/DVD by Image Entertainment

General Reference Books

❑ *Windows on Nature: Animals of the Polar Region*
❑ *Polar Wildlife* by Joshua Morris
❑ *A for Antarctica* by Jonathan Chester
❑ *Antarctica* (A True Book) by David Petersen
❑ *Polar Lands* by Christopher Green
❑ *Take a Trip to Antarctic* by Keith Lye
❑ *Antarctica* (A New True Book) by Lynn M. Stone
❑ *Polar Exploration* by Martyn Bramwell
❑ Pages 72-75 *Our Father's World*

Literature

❑ The Strange Case of Moody, Watch, and Spy – Your Story Hour audio series

Science

❑ Penguins
  • *Plenty of Penguins* by Sonia W. Black
  • *Baby Penguin* – Baby Animal Stories
  • http://encarta.msn.com/find/MediaList.asp?pg=6&mod=2&ti=761576981

Notebook Suggestions:

1. What is a colony of penguins called?
2. Explain how penguins use their wings.
3. How long can penguins stay under water?
4. Describe how penguin parents keep their egg and baby warm.

Vocabulary

ice	iceberg	floe
ice sheet	icecap	snow
icebreaker	glacier	

Music/Art/Projects

1. The Antarctic is "owned" internationally and does not have a flag of its own. Let the child design a flag.

2. Make penguin suits.

    Instructions: Paint a paper grocery sack black. Allow to dry completely. Cut out a belly from white paper and glue to sack. Cut a hole big enough for the child's head. To make flippers, cut rounded flaps from the side panels. Cut a beak from orange construction paper and glue to a black mask (can be found at a party supply store).

Internet Resources

❑ http://encarta.msn.com/find/MediaList.asp?pg=6&mod=2&ti=761565002

Bible

❑ Snow
  • Job 24:19
  • Psalm 147:15 & 16
  • Job 38:22 &23
  • Isaiah 1:18-20
  • Job 37:6
  • Isaiah 55:10
❑ Ice
  • Job 38:30
  • Job 6:15 & 16

Word Search

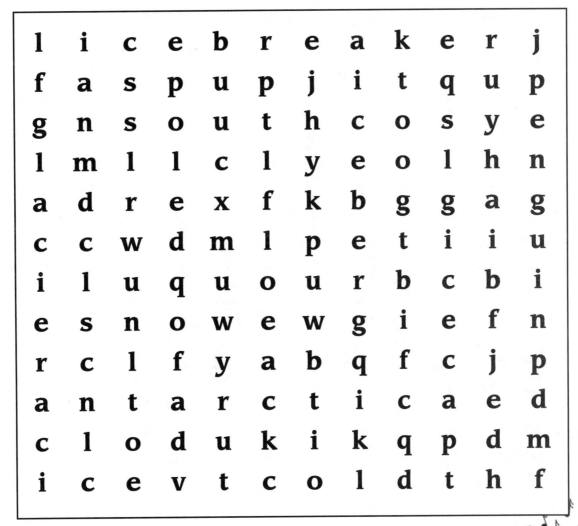

```
l i c e b r e a k e r j
f a s p u p j i t q u p
g n s o u t h c o s y e
g l m l l c l y e o l h n
a d r e x f k b g g a g
c c w d m l p e t i i u
i l u q u o u r b c b i
e s n o w e w g i e f n
r c l f y a b q f c j p
a n t a r c t i c a e d
c l o d u k i k q p d m
i c e v t c o l d t h f
```

Antarctica
cold
floe
glacier
ice
iceberg

icebreaker
icecap
penguin
pole
snow
south

The Arctic

This is God's view from the top of the world.

What 3 continents have land near the North Pole?

1. _____

2. _____

3. _____

The Arctic

The bitterly cold region known as the Arctic lies north of the Arctic Circle. Portions of Canada, the United States, Greenland, Norway, Sweden, Finland, and Russia lie within the Arctic Circle. The Arctic Ocean is the world's smallest ocean covering an area of nearly 5 million square miles. The average depth is 3,675 feet. The greatest known depth is 17,880 feet. Near the center of the Arctic Ocean is the North Pole.

People/History

❑ Robert Peary & Matthew Henson
  • Page 136 – *Usborne Book of Famous Lives*
  • *Matthew Henson: Arctic Explorer* by Jeri Ferris

❑ Henry Hudson
  • Page 122 – *Usborne Book of Famous Lives*
  • *Henry Hudson* by Ruth Harley

General Reference Books

❑ *Windows on Nature: Animals of the Polar Region*
❑ *Polar Wildlife* by Joshua Morris
❑ *Polar Lands* by Christopher Green
❑ *Arctic Babies* by Kathy Darling
❑ *Arctic Lands* edited by Henry Pluckrose
❑ *Polar Exploration* by Martyn Bramwell
❑ *Arctic Spring* by Sue Vyner
❑ Arctic Kingdom: Life At The Edge – National Geographic Video

Literature

❑ *Little Polar Bear, Take Me Home* by Hans de Beer

Science

❑ Wolverine
  • Page 64 – *Special Wonders of the Wild Kingdom*
  • http://encarta.msn.com/find/MediaMax.asp?pg=3&ti=761557917&idx=461518902
  Notebook Suggestions:
    1. The wolverine is the largest member of which family?
    2. Describe what is special about wolverine fur.
    3. Explain why wolverines are thought of as fierce animals.
    4. When are wolverine cubs born?

❑ Arctic fox
  • *Arctic Foxes* by Downs Matthews
  • http://encarta.msn.com/find/MediaMax.asp?pg=3&ti=761565627&idx=461517524

Notebook Suggestions:

1. What color is the arctic fox in summer?
2. What color is the arctic fox in winter?
3. Why does it change color?
4. Explain how the arctic fox conserves heat.
5. What happens to the arctic fox at –94 F?

❑ Seals

- Cousteau Society: *Seals*
- *Harp Seal Pups* by Downs Matthews
- http://encarta.msn.com/find/MediaList.asp?pg=6&mod=2&ti=761564979

Notebook Suggestions:

1. Explain what move seals quickly through water.
2. What keep seals warm?
3. Which kind of seal moves easily on land?
4. What color are seal pups?
5. Why are the pups this color?

❑ Orca (Killer Whale)

- Lesson 26B - *Considering God's Creation*
- *Killer Whales* by Dorothy Hinshaw Patent
- *Killer Whales* by Mark Carwardine
- http://encarta.msn.com/find/MediaMax.asp?pg=3&ti=761579745&idx=461539265 (membership required)

Notebook Suggestions:

1. Is an orca a dolphin or a whale?
2. What color is an orca?
3. How much food does an orca eat each day?
4. Orcas travel family groups called _____ .
5. Explain how orcas communicate.

❑ Snowy Owls

- *Snowy Owl at Home on the Tundra*
- Snowy Owl – *Baby Animal Stories*
- http://encarta.msn.com/find/MediaMax.asp?pg=3&ti=761554738&idx=461530336

Notebook Suggestions:

1. Describe what snowy owls have covering their feet and legs.
2. Explain why owls must turn their heads to look around.
3. What is the main food source for snowy owls?
4. Describe where snowy owls build their nests.

❑ Polar bears
  • *Polar Bears* by Marcia S. Freeman
  • *Polar Bear Cubs* by Downs Matthews
  • Page 48 & 49 – *Special Wonders of the Wild Kingdom*
  • *Polar Bear Alert* – National Geographic Explorer Video
  • http://encarta.msn.com/find/MediaMax.asp?pg=3&ti=761579749&idx=461530442 (membership required)

  Notebook Suggestions:
  1. What features keep polar bears warm?
  2. Wh do polar bears have webbed toes?
  3. How long do polar bear cubs stay with their mother?

Vocabulary

ice	ice sheet	floe
snow	icecap	tundra
icebreaker	igloo	iceberg
glacier		

Music/Art/Projects

❑ Make a model igloo.
  1. In a bowl, mix 3 cups flour, 1 1/2 cups salt, and 3/4 cup water. Gradually add another 3/4 cup water. Knead until a ball is formed.
  2. Press dough into two ice cube trays. Freeze at least two hours. Remove blocks.
  3. On a piece of paper draw or trace a circle 4 inches across. Fit the blocks on their sides along the circle with the small ends facing inward.
  4. Igloos are built on a spiral. Use a ruler to cut the first two blocks at a slant, forming a ramp. Continue to place blocks around the circle and up the ramp in a spiral. Trim blocks to fit as they close in to form a dome.
  5. When the dome is finished, use trimmings to fill in the gaps. Cut a door at the base.
  6. To harden, bake in a 125° oven.

Internet Resources

❑ For more information on igloos go to: http://indy4.fdl.cc.mn.us/~isk/maps/houses/igloo.html
❑ http://encarta.msn.com/find/MediaList.asp?pg=6&mod=2&ti=761577860

Bible

❑ Refer to Bible references for Antarctica.

Word Search

```
w o l v e r i n e s t
b h d z p w l l r n s
a f r s e a l v f o n
r o m o f f u e o w l
c x m p g c b a c y f
t u n o r t h p o y d
i y z l e x r v r r i
c p s e b m s i c f g
t h p o t b w z a o l
p o l a r b e a r n o
j y m q t u n d r a o
```

Arctic
fox
igloo
north
orca
owl

polar bear
pole
seal
snowy
tundra
wolverine

120

MAZE CRAZE

FINISH

North America

North America is the third largest continent with the fourth largest population. It is bordered by the Atlantic Ocean on the east and the Pacific Ocean on the west. The highest point is Mt. McKinley in Alaska (20,320 ft.) and the lowest point is at Death Valley, California where the elevation is 282 ft. below sea level. North America's vast natural resources and rich mineral reserves have produced two of the largest manufacturing countries in the world.

North America

Color each country you study.

Artic Ocean

Greenland

USA

Labrador Sea

Canada

Hudson
Bay

Lake
Superior
Lake
Huron
Lake
Michigan
Lake
Ontario
Lake Erie

United States of America

Atlantic Ocean

Mexico

Gulf
of
Mexico

Bahamas

Cuba

Dom. Rep.

Jamacia

Haiti

Belize
Guatemala
Honduras
El Salvador
Nicaragua

Caribbean Sea

Pacific Ocean

Costa Rica

Panama

© 2003 Geography Matters

ad maiorem Dei gloriam!

Canada

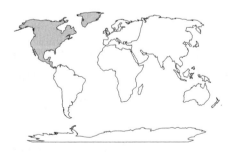

Population_____

Capital City_____

Religion_____

Type of Government_____

Currency_____

Language_____

What are the people called?_____

Canada

Canada has two official languages, English and French. Both languages are used by the federal government and are seen throughout Canada.

Canada's wealth comes from manufacturing and natural resources, which include uranium, zinc, gold, and oil. Canada is also one of the world's largest producers of wheat. Almost half of Canada's land is covered with forest. The forests are made up of many kinds of trees, including cedar, hemlock, fir, and pine. Canada is one of the leading lumber-producing countries in the world. The numerous lakes and rivers provide a natural transportation system for logs and other products of the industry. Christmas tree farms in Nova Scotia, Quebec, and New Brunswick send trees as far away as South America.

Many fishing villages are tucked back into sheltered coves of Canada's coast. The fishing fleets of Nova Scotia are among the finest in the world. Many of the men who do not go out in the fishing vessels work in the canneries and processing plants. Here they help to prepare the haddock, codfish, and tuna for shipment all over the world.

The national parks of Canada, especially those in the Canadian Rocky Mountains, are among the most beautiful in the world. Along the border between the United States and Canada are playgrounds which the two countries share. Canada's Waterton Lakes National Park and Glacier National Park of the United States join each other. They make one large park which is called the International Peace Park. There is a marker on a big stone at the border which tells of the friendship between the two countries. The words on the marker are: "To God in His Glory, we two nations dedicate this garden and pledge ourselves that as long as men shall live, we will not take up arms against one another."

Canadian boys love to play the fast, exciting game of ice hockey. This game is so popular in Canada that it is the countries national sport. Girls rarely play the game because it is so rough, but they do enjoy figure skating and many other winter sports.

People/History

❑ Royal Canadian Mounted Police (Mounties)

The Mounties are as much a symbol of Canada as the maple leaf. The Mounties were originally recruited in 1873 to prevent bloodshed between whiskey traders and Native People in the Northwest Territories. Riding horseback, they brought law and order to the expanding Canadian frontier.

The Mounties' red coats were a symbol of peace, the color was chosen because the Native People equated red with justice and fair dealing. The broad-brimmed hats were adopted by 1900 because they offered protection from the sun. Today, the Mounties wear the red coat for dress and ceremonial occasions, including parades.

General Reference Books

- ❑ *North America* (A True Book) by David Petersen
- ❑ *A is for the Americas* by Cynthia Chin-Lee & Terri de la Pena
- ❑ Pages 8-30 *Our Father's World*
- ❑ *Take a Trip to Canada* by Keith Lye
- ❑ *Count Your Way Through Canada* by Jim Haskins
- ❑ *An Eskimo Family*
- ❑ Page 23, *Children Just Like Me* by Barnabas & Anabel Kindersley

Literature

- ❑ Chapters 28 & 29 – *Missionary Stories with the Millers*
- ❑ *I Know an Old Lady Who Swallowed a Fly* retold by Nadine Bernard Westcott
- ❑ *Very Last First Time* by Jan Andrews
- ❑ *Chester's Barn* by Lindee Climo
- ❑ *A Regular Rolling Noah* by George Ella Lyon
- ❑ *Prairie Boys Winter* by William Kurelek
- ❑ *The Flight of the Union* by Tehla White

Science

- ❑ Moose
  - Pages 44 - *Special Wonders of the Wild Kingdom*
  - http://encarta.msn.com/find/MediaMax.asp?pg=3&ti=761563637&idx=461517536

  Notebook Suggestions:
  1. How many points do moose antlers have?
  2. How fast can a moose trot?
  3. How fast can a moose run?
  4. How much does a baby moose weigh?

- ❑ Lesson 23 – *Considering God's Creation* - Animal Food Chain

- ❑ Play "Into the Forest" by Ampersand Press.

- ❑ If you read *Very Last First Time*, discuss dressing for weather conditions.

- ❑ If you read *Chester's Barn*, check out some books on different farm animals from the library or read ones already on your shelf.
  - *Farm Animals* (A New True Book) by Karen Jacobsen

Notebook Suggestions:

1. Explain how pigs cool themselves.
2. What do pigs use to find and dig up food?
3. How many parts does a cow's stomach have?
4. List what foods are made from cow's milk.
5. Describe how chickens use their strong claws.

❑ If you read *A Regular Rolling Noah,* continue with study of farm animals and/or read a book about trains.

Vocabulary

lumberjack	Northwest Passage	snowmobile
Acadian	voyageur	micmac

Music/Art/Projects

1. Color or make the flag of Canada.

   Red represents the sacrifice made by Canadians during World War I. White represents the snowy north of Canada. The maple leaf is a Canadian symbol because of the abundance of maple trees in Canada.

2. Color or make a map of Canada.

3. Play or watch hockey.

4. Celebrate Canada's winter climate with a snowball fight. If no snow is available, use wadded paper or foam balls.

5. Make Canadian Baked Blueberry Dessert

16 oz. fresh or frozen blueberries	1 1/2 tablespoons lemon juice
2 teaspoons cornstarch	1 cup packed brown sugar
1/2 cup flour	2/3 cup quick oats
1/4 teaspoon salt	1/3 cup margarine

   Directions: In an ungreased casserole dish, toss blueberries with lemon juice. In a separate bowl, mix cornstarch with 1/2 cup of brown sugar. Stir into blueberries. Mix flour, oats, 1/2 cup sugar, and salt. Cut in margarine with fork and sprinkle over blueberries. Bake uncovered at 350° for 45 minutes. Serve warm with vanilla ice cream.

6. Make Habitant Pea Soup

    This is a French-Canadian dish adapted from the traditional food carried by voyageurs on their long trips.

1 1/4 lbs. dried peas	2 diced onions
1/2 lb. Salt pork	3 bay leaves
11 cups water	1 tsp. salt
1/2 cup chopped celery	1 tsp. savory
1/4 cup chopped parsley	

    Directions: Wash and drain peas. Put them in a soup pot with water. Boil for two minutes; remove from heat and cool two hours. Add remaining ingredients. Bring soup to a boil again; reduce heat and simmer two hours.

7. Make snowshoes.
    - Have child stand on poster board or cardboard.
    - Trace around each foot. Measure an oval 6 inches larger than the outline of each foot.
    - Cut out cardboard or poster board.
    - Have the child stand on the cut out boards.
    - Make markings for three to five holes on each side of each foot.
    - Punch holes at markings.
    - Lace the holes with a long piece of yarn or twine.
    - Tie in place over child's shoes.

8. Make a Canadian forest scene.
    - To make tree trunks, cut 1 inch wide strips of brown construction paper.
    - Glue onto white construction paper.
    - Tear small pieces of green tissue paper or construction paper.
    - Glue onto white construction paper in an overlapping pattern to make the leaves on the trees.

9. Make a maple leaf mobile - option 1.
    - Color copies of the maple leaf pattern on page 131 red or use them as a pattern to trace onto red paper or photocopy onto red paper.
    - Cut out leaves.
    - Poke a small hole in the top of each leaf.
    - Thread string through the hole and tie onto a clothes hanger at varying lengths.

10. Make a maple leaf mobile - option 2.
    - Color four copies of the medium size maple leaf pattern red. Cut out.
    - Trace a CD onto white paper four times. Cut out the four circles.
    - Glue a red maple leaf to the center of each white circle. Allow to dry.

- Gently fold each circle in half so that the fold runs through the leaf vertically.
- Match the spines of two circles side-by-side. Glue edges together. Repeat with other two circles.
- Allow to dry.
- Tie a knot on one end of an 18" piece of yarn.
- Glue the two halves together with the yarn running through the middle. Allow to dry.

Internet Resources

❑ http://encarta.msn.com/find/MediaList.asp?pg=6&mod=2&ti=761563379

Bible

❑ Chapter 28 - *Missionary Stories with the Millers*
- I Thessalonians 5:18
- 2 Thessalonians 3:3
- Psalm 46:1
- Joshua 1:5

❑ Chapter 29 - *Missionary Stories with the Millers*
- Acts 27

❑ *Very Last First Time*
- Isaiah 41:10
- Psalm 27:1

❑ *A Prairie Boy's Winter*
- Ecclesiastes 3:1-8

❑ Farm Animals
- Mark 5:11-13
- Luke 15:11-32
- Matthew 7:6
- 2 Peter 2:22
- Genesis 30:31-43
- Exodus 26:7

Flag of Canada

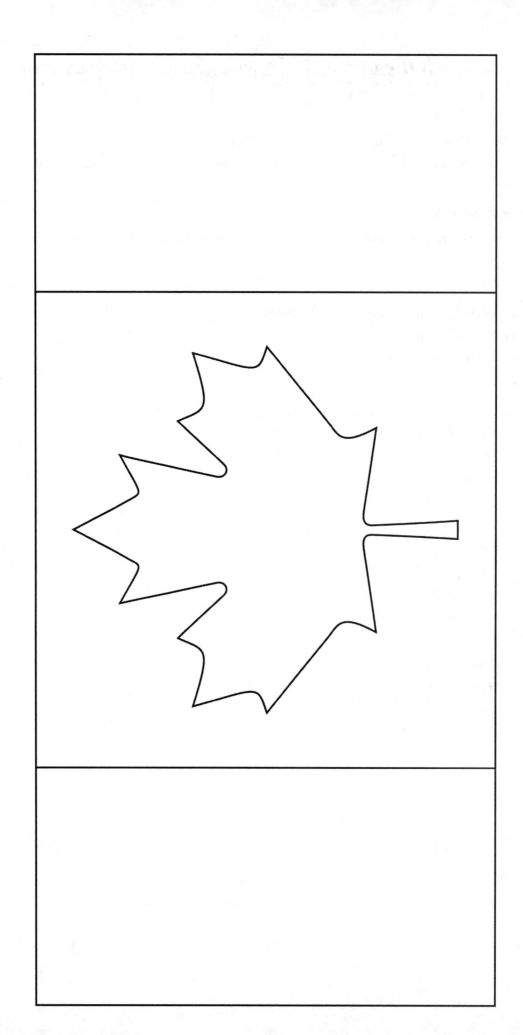

Maple Leaf Pattern

131

FIND THE TWINS

Which two are exactly alike?

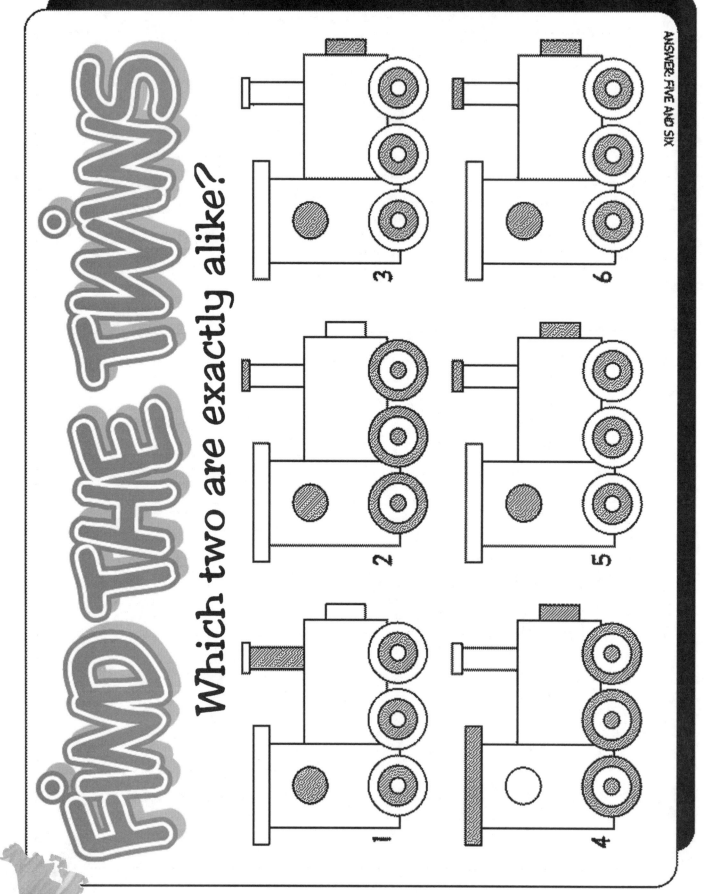

ANSWER: FIVE AND SIX

132

MAZE CRAZE

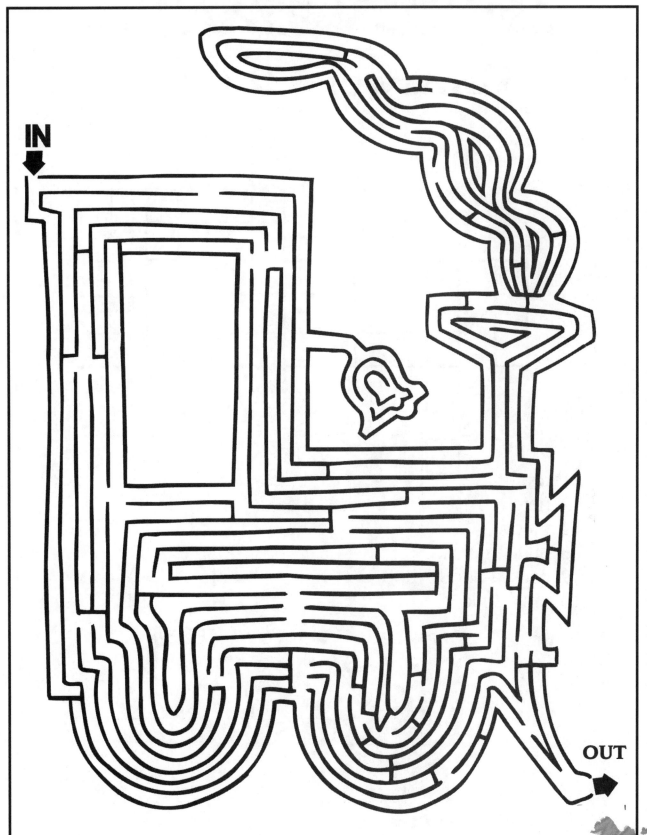

IN

OUT

Crossword Puzzle

Across

2 neigh

4 meow

6 quack-quack

9 cock-a-doodle-doo

10 oink-oink

Down

1 cluck-cluck

3 baa

4 moo

5 gobble-gobble

7 woof-woof

Word Search

```
c  g  d  u  c  k  n  n  p  g
o  p  o  c  u  w  m  p  i  h
w  o  g  o  o  s  e  i  g  g
s  c  f  f  i  y  a  r  o  l
w  h  o  r  s  e  q  o  b  s
z  i  v  q  z  a  y  o  n  h
g  c  m  u  g  h  a  s  t  e
c  k  s  s  o  u  r  t  m  e
c  e  a  x  a  i  w  e  k  p
s  n  f  t  t  w  d  r  f  t
```

chicken	goat	pig
cow	goose	rooster
duck	horse	sheep

135

The United States of America

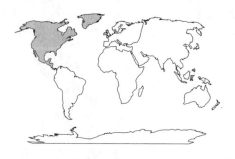

136

Population_____

Capital City_____

Religion_____

Type of Government_____

Currency_____

Language_____

What are the people called?_____

United States

The United States is often called a "melting pot" because people of different races and from different lands make their home here. Many of the states have names of Native American origin. For example, Iowa comes from the word "Ayuhwa" meaning "sleepy one". Mississippi comes from "misi" meaning "big" and "sipi" meaning "river". Wisconsin acquired its name from the word "wishkonsing" which means "place of the beaver". The "D.C." in Washington D.C., stand for District of Columbia. The district was named for Christopher Columbus, and the city for George Washington.

Virginia has been the home state to the largest number of presidents, eight in all. The flag of the United States is the only flag to be the subject of the country's national anthem. The national bird was almost a turkey. Benjamin Franklin proposed that the turkey be the national bird. In 1782, it was decided that the bald eagle, which is unique to North America, would be the choice.

People/History

There are too many historical figures to be covered in this section. We chose to cover our current president and three past presidents.

- ❏ George Washington
    - *George Washington* by Garnet Jackson
    - *Buttons for General Washington* by Peter & Connie Roop
    - *Meet George Washington* by Heilbrendt
    - *A Picture Book of George Washington* by David A. Adler
    - *Phoebe the Spy* by Judith Berry Griffin

- ❏ Thomas Jefferson
    - *Meet Thomas Jefferson* by Marvin Barrett
    - *Young Thomas Jefferson* by Francene Sabin
    - *A Picture Book of Thomas Jefferson* by David A. Adler

- ❏ Abraham Lincoln
    - *Abe Lincoln's Hat* by Martha Brenner
    - *Just a Few Words, Mr. Lincoln* by Jean Fritz
    - *Meet Abraham Lincoln* by Carey
    - *A Picture Book of Abraham Lincoln* by David A. Adler

- ❏ Geroge W. Bush (or current president)
    - www.whitehouse.gov

- ❏ Native American Indians
    - *Indians* (A New True Book) by Teri Martini

❑ American Flag
  • *Our Flag* by Leslie Waller
  • *The American Flag* (A True Book) by Patricia Ryon Quiri
  • *Red, White, and Blue* by John Herman
  • *The Star Spangled Banner* by Peter Spier

General Reference Books

❑ *Welcome to the Sea of Sand* by Jane Yolen
❑ *Yosemite National Park* (A New True Book) by David Petersen
❑ Pages 18-22, *Children Just Like Me* by Barnabas & Anabel Kindersley
❑ *Mount Rushmore* by Thomas S. Owens
❑ *Alaska* by Joyce Johnston
❑ *Rocky Mountain National Park* (A New True Book) by David Petersen
❑ *First Facts About the United States* by David L. Steinecker
❑ *America's Forests* by Frank Staub

Literature

❑ Chapters 22 & 25, & the Introduction – *Missionary Stories with the Millers*
❑ Chapters 5, 9, 14, & 20 – *Heaven's Heroes*
❑ The Cranberry series by Wende & Harry Devlin are wonderful books set in New England. There are several titles in the series – *Cranberry Thanksgiving, Cranberry Valentine, Cranberry Easter*, etc. Choose the one that is set in the season you are in. Plan ahead and make the recipe at the end of the book.
❑ *When I Was Young in the Mountains* by Cynthia Rylant
❑ *Lentil* by Robert McClosky
❑ *Yankee Doodle* by Edward Bangs
❑ *The Story of the Statue of Liberty* by Betsy & Giulio Maestro
❑ *Copper Lady* by Alice Ross
❑ *Maybelle, the Cable Car* by Virginia Lee Burton
❑ *The Snow Walker* by Margaret K. Wetterer
❑ *Chang's Paper Pony* by Eleanor Coerr
❑ *My Great Aunt Arizona* by Gloria Houston
❑ *Only Opal* by Barbara Cooney
❑ *First Flight* by George Shea
❑ *Countdown to Flight* by Steve Englehart
❑ *They Were Strong and Good* by Robert Lawson
❑ *Oxcart Man* by Barbara Cooney

Science

❑   Buffalo

- *Buffalo* (A New True Book) by Emilie U. Lepthien
- Page 10 - *Special Wonders of the Wild Kingdom*
- Lesson 26A – *Considering God's Creation*
- http://encarta.msn.com/find/MediaMax.asp?pg=3&ti=761570386&idx=461517555

Notebook Suggestions:

1. What is another name for buffalo?
2. Why were buffalo so important to the plains Indians?
3. How far can buffalo see, smell, and hear?
4. How fast can buffalo run?

❑   Rabbits

- Little Jackrabbit – *Baby Animal Stories*
- *Cottontail Rabbits* by Kristin Ellerbusch Gallagher
- *Diary of a Rabbit* by Lilo Hess
- *Rabbits, Rabbits, & More Rabbits* by Gail Gibbons
- http://encarta.msn.com/find/MediaList.asp?pg=6&mod=2&ti=761568905

Notebook Suggestions:

1. Rabbits are members of which family?
2. Never pick up a rabbit by its _____ .

❑   Beaver

- Page 14 - *Special Wonders of the Wild Kingdom*
- *Beavers* by Deborah Hodge

Notebook Suggestions:

1. What are beaver's cutting teeth called?
2. How does a beaver use its tail?
3. What is a beaver home called?
4. What are the homes made of?
5. Is the entrance above water or below water?

❑   Otters

- *Sea Otters* by Evelyn Shaw
- *Playful Slider* by Barbara Juster Esbensen
- *Oopsie Otter* by Suzanne Tate
- http://encarta.msn.com/find/MediaMax.asp?pg=3&ti=761555318&idx=461525572 (membership required)

Notebook Suggestions:

1. Where do otters spend most of their time?
2. What kind of feet do otters have?
3. How do otters communicate with each other?

❑ Squirrels

• *The Squirrel and the Nut* – God is Good Series by Rod & Staff

• http://encarta.msn.com/find/MediaList.asp?pg=6&mod=2&ti=761556073

Notebook Suggestions:

1. How do squirrels use their claws and tails?

2. Why do squirrels store nuts in the fall?

❑ Skunk

• Page 58 - *Special Wonders of the Wild Kingdom*

• *Skunks* by Sandra Lee

• http://encarta.msn.com/find/MediaMax.asp?pg=3&ti=761577193&idx=461517409

Notebook Suggestions:

1. What two things are skunks known for?

2. A skunk is about the same size as a small _____ .

3. Where are skunks found?

4. How are skunks helpful to people?

5. How does a skunk warn an animal to go away?

❑ Raccoons

• Page 54 - *Special Wonders of the Wild Kingdom*

• *Baby Raccoon* by Beth Spanjian

• *Clever Raccoons* by Kristin L. Nelson

• http://encarta.msn.com/find/MediaMax.asp?pg=3&ti=761568954&idx=461546811

Notebook Suggestions:

1. Why do raccoons eat more during the fall?

2. What is a baby raccoon called?

3. Why do raccoons wash their food?

❑ Porcupine

• Page 50 - Special Wonders of the Wild Kingdom

• http://encarta.msn.com/find/MediaMax.asp?pg=3&ti=761551922&idx=461517271

Notebook Suggestions

1. What are quills?

2. Why don't porcupines attack other animals?

3. When do porcupines sleep?

4. What does a porcupine do when it feels threatened?

❑ Cougars
  • *Baby Cougar* by Beth Spanjian
  • http://encarta.msn.com/find/MediaMax.asp?pg=3&ti=761572249&idx=461516872
  Notebook Suggestions:
    1. What is another name for cougar?
    2. Name two different climates cougars can live in?

❑ Prairie Dogs
  • *Prairie Dog at Home on the Range* by Sarah Toast
  • Prairie Dog – *Baby Animal Stories*
  • http://encarta.msn.com/find/MediaMax.asp?pg=3&ti=761561750&idx=461514788
  Notebook Suggestions:
    1. A prairie dog is really a _____.
    2. What kind of flower seeds do prairie dogs like?

❑ Manatees
  • *I Wonder if Sea Cows Give Milk* by Annabelle Donati
  • *Manatee Winter* by Kathleen Weidner Zoehfeld
  • *A Safe Home for Manatees* by Priscilla Belz Jenkins
  • *A Manatee Morning* by Jim Arnosky
  • http://encarta.msn.com/find/MediaMax.asp?pg=3&ti=761575938&idx=461547789
  Notebook Suggestions:
    1. What is another name for the manatee?
    2. What color are manatees?
    3. How big are manatees?

❑ General Animal Books
  • *Armadillos Sleep in Dugouts and Other Places Animals Live* by Pam Munoz Ryan
  • *Animal Babies* Illustrated by Fiammetta Dogi
  • *God Made the Animals* – God is Good Series by Rod & Staff
  • *Animal Homes* – First Little Golden Book
  • *Big Tracks, Little Tracks* by Millicent E. Selsam
  • *America: Land of Wildlife* by Karen Jensen
  • *Wonders of the Forest* by Francene Sabin

Vocabulary

herbivore	predator	freedom
carnivore	prey	republic
omnivore	liberty	melting pot

141

Music/Art/Projects

1. Color or make the flag of the United States.

    The thirteen stripes represent the thirteen original colonies. The fifty white stars stand for the fifty states that make up the union.

2. Color or make a map of the United States.

3. Put together a puzzle of the United States.

    We prefer the puzzles that have each state as an individual piece. Most teacher supply stores carry them as well as Timberdoodle (360-426-0672 or www.timberdoodle.com) and Hands On and Beyond(1-888-20- LEARN or www.HandsOnAndBeyond.com).

4. Try recipes from regions of the country other than your own.

5. Listen to The Stories of Foster & Sousa – Music Masters Series

6. Make American flag graham crackers.

    Ingredients: graham crackers, white chocolate chips (for the stars), tubes of red, white, and blue frosting or white frosting with red and blue sugar sprinkles for stripes and blue field.

7. Listen to the "Star Spangled Banner" and other patriotic music.

    Wee Sing America is a good source of patriotic music for children.

8. Find out the origin of your state's name.

9. Many communities have wildlife refuges or nature centers. Visit one in your area.

10. Make a star mobile - option 1.
    • Color copies of the star pattern on page 152 or use them as a pattern to trace onto red, white and blue paper or photocopy onto red, white and blue paper.
    • Cut out stars.
    • Poke a small hole in the top of each star.
    • Thread string through the hole and tie onto a clothes hanger at varying lengths.

11. Make a star mobile - option 2.
    • Copy or trace the largest star pattern onto white paper four times. Cut out four stars.
    • Trace a cd onto red paper two times and onto blue paper two times. Cut out the four circles.
    • Glue a white star to the center of each red and blue circle. Allow to dry.
    • Gently fold each circle in half so that the fold runs vertically through the tip of each star.
    • Match the spines of two circles side-by-side. Glue edges together. Repeat with other two circles.

- Allow to dry.
- Tie a knot on one end of an 18" piece of yarn.
- Glue the two halves together with the yarn running through the middle. Allow to dry.

Internet Resources

❑ www.whitehouse.gov - Has a wonderful section for children.

❑ http://encarta.msn.com/find/MediaList.asp?pg=6&mod=2&ti=1741500822

Bible

❑ Chapter 22 - *Missionary Stories with the Millers*
- Galations 6:2

❑ Chapter 25 - *Missionary Stories with the Millers*
- Matthew 13:1-9, 18-23
- Luke 12:20
- Matthew 25:13

❑ *When I Was Young in the Mountains*
- John 3:16
- John 1:32-34
- Matthew 3:13-17
- Acts 8:26-40
- Matthew 28:19-20

❑ *Lentil*
- Proverbs 18:16
- Genesis 4:2-15
- John 21:20-23

❑ *Maybelle, the Cable Car*
- Deuteronomy 32:7

❑ *Chang's Paper Pony*
- Proverbs 17:5
- Leviticus 19:35-36
- Deuteronomy 25:15
- Proverbs 11:1
- Revelation 21:21b

❑ *The Snow Walker*
- Matthew 25:35a
- Job 29:15-16

❑ *They Were Strong and Good*
- Genesis 12:3b
- Genesis 25:12-26
- Matthew 1:1-17

❑ *Oxcart Man*
- Acts 20:34
- Ephesians 4:28b
- Proverbs 6:6-8

❑ Rabbit
- Leviticus 11:6

Flag of the United States of America

U.S.A Time Zones

Alaska

Mountain

Pacific

Central

Eastern

© 2003 Geography Matters

ad maiorem Dei gloriam!

Color your state.

What are the people from your state called? _____

What is your state nickname?   _____

In which time zone do you live? _____

MAZE

Find the capital of Kansas.

Start Here

Topeka ★

U.S.A.

MAZE CRAZE

START

FINISH

Crossword Puzzle

Across

6 The longest river in the U.S.
8 This man is credited with discovering America.
11 The Aloha State.
12 This President ended slavery.
14 The Peach State.
15 The Bluegrass State.

Down

1 The Grand Canyon State.
2 The first President.
3 Old Dominion.
4 The Show-me State.
5 The Lone Star State.
7 The first state to join the Union.
9 The number of states in the U.S.
10 The Sunshine State.
13 The Garden State.

Word Search

```
x m n e a t w n p l x u a
w a s h i n g t o n d c m
o x c w k e y j g x u q e
s t r i p e s m l e o s r
t c u h j t w f b k h c i
a v z z r d h g l x t r c
r x a f e x i m u p w j a
s f g h d p t w e l n v f
v f c s k s e o u x o x t
g i d f r e e d o m a o z
f f a i e v j j h c p o z
s t a t e s v x x m l r f
f y g g c n l i b e r t y
```

America   stars
blue   states
fifty   stripes
freedom   Washington DC
liberty   white
red

Word Search

```
p  z  t  p  a  m  a  n  a  t  e  e
r  h  b  e  a  v  e  r  i  l  c  z
a  f  a  z  k  e  s  o  b  b  i  o
i  x  y  b  f  o  q  m  u  y  q  t
r  d  g  b  j  e  u  y  f  e  r  t
i  q  b  m  q  p  i  q  f  r  a  e
e  c  o  u  g  a  r  w  a  j  c  r
d  t  h  m  f  b  r  l  l  l  c  x
o  s  k  u  n  k  e  r  o  t  o  d
g  o  l  x  o  t  l  w  u  l  o  s
t  r  f  p  o  r  c  u  p  i  n  e
r  a  b  b  i  t  t  r  z  p  j  h
```

beaver
buffalo
cougar
manatee
otter
porcupine

prairie dog
rabbit
raccoon
skunk
squirrel

 **Notes:**

Mexico

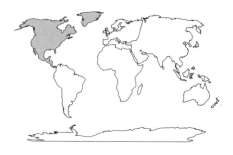

152

Population_____

Capital City_____

Religion_____

Type of Government_____

Currency_____

Language_____

What are the people called?_____

Mexico

Mexico's land is composed of mountains, deserts, and tropical zones. There are three mountain ranges; all of them are named Sierra Madre. Farmers make up approximately one-third of the Mexican population. Today, many farmers are trying to immigrate to the United States in search of work because they can't find work in Mexico's overcrowded cities.

Some Mexicans use adobe, a mixture of wet clay and straw, to build houses. On hot days, adobe walls keep temperatures cooler inside the house. The tortilla, Mexico's most famous bread, is an unleavened corn or flour cake. Preparation includes soaking corn kernels in limewater until they are soft enough to grind, then adding water a little at a time, to make the dough. The dough is rolled until it is very thin and then it is baked.

People/History

❑ *Charro: The Mexican Cowboy* by George Ancona

❑ Pages 120 & 121 – *Usborne Book of Famous Lives*

❑ Benito Juarez
  • *Benito Juarez: Hero of Modern Mexico* by Rae Bains

General Reference Books

❑ Pages 105-111 *Our Father's World*
❑ *A Family in Mexico* by Tom Moran
❑ *Count Your Way Through Mexico* by J. Haskins
❑ *Passport to Mexico* by C. Irizarry
❑ *Mexico* (A New True Book) by K. Jocobsen
❑ *Mexico* by Kate A. Furlong
❑ *Inside Mexico* by Ian James
❑ *Mexico* (A True Book) by Ann Heinrichs
❑ *Cinco de Mayo* by Lola M. Schaefer
❑ Pages 16 & 17, *Children Just Like Me* by Barnabas & Anabel Kindersley

Literature

❑ *Hill of Fire* by Thomas P. Lewis
❑ *Happy Days with Pablo and Juanita* by Evelyn Hege (Rod & Staff)
❑ Chapters 10 & 16 – *Missionary Stories with the Millers*

Science

❑ Volcanoes
  • *Volcanoes* by Franklyn M. Branley
  • *Volcanoes! Mountains of Fire* by Eric Arnold

- *Why Do Volcanoes Blow Their Tops?* by Melvin & Gilda Berger
- *Volcanoes (A New True Book)* by Helen J. Challand
- http://encarta.msn.com/find/MediaList.asp?pg=3&ti=761570122

Vocabulary

fiesta	cactus	pinata
charro	sombrero	volcano
burro	adobe	tortilla

Music/Art/Projects

1. Color the flag of Mexico.

    The flag has three vertical bands, green, white, and red. Green stands for independence, white for religion, and red for union. The coat of arms in the center illustrates an Aztec legend that explains the founding of Mexico City.

2. Color or make a map of Mexico.

3. Make tissue paper flowers.
    - Take six sheets of tissue paper (solid or multi-colored) and fold them like a fan.
    - Cut the folded paper in half.
    - Take one half at a time, still folded, and trim both ends into a broad point.
    - Bend a twist tie around the center.
    - Carefully separate each layer of the folded paper.
    - If necessary, tape the ends together so that the flower will stay open.

4. Use clay to make an adobe house. Let the house dry in the sun.

5. Eat Mexican food.
    - *Food in Mexico* by Pablo Gomez
    - *Cooking the Mexican Way* by Rosa Coronado

6. Learn to count to ten in Spanish.

uno – one	seis - six
dos – two	siete - seven
tres – three	ocho - eight
cuatro – four	nueve - nine
cinco – five	dies - ten

Internet Resources

- www.mexonline.com
- http://encarta.msn.com/find/MediaList.asp?pg=6&mod=2&ti=761576758

Bible

- *Volcano*
  - Psalm 104:32
- Donkey
  - Matthew 21:1-7
  - Proverbs 26:3
  - Genesis 22:3
  - Job 39:5-8
  - Job 6:5
- Chapter 10 - *Missionary Stories with the Millers*
  - Acts 16:16-40

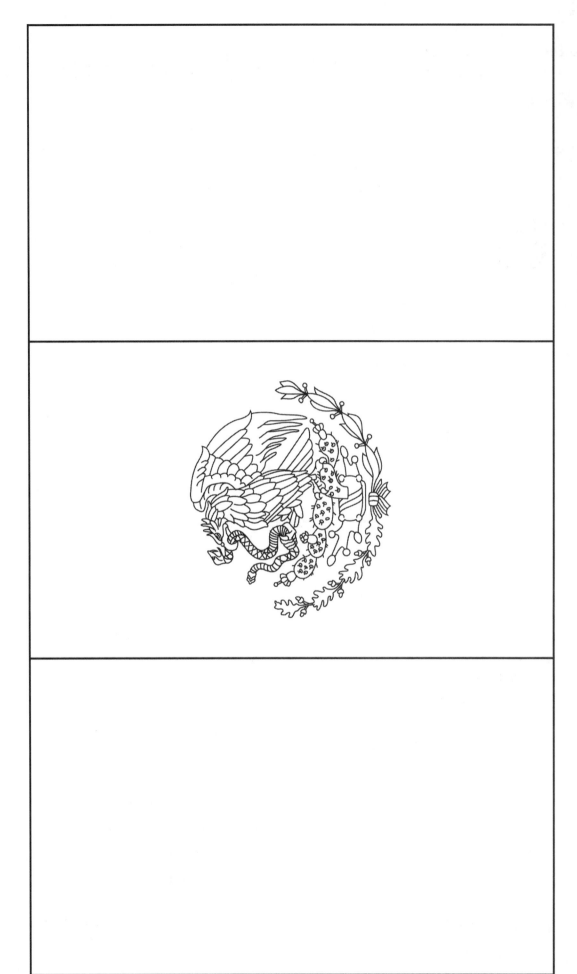

Flag of Mexico

Crossword Puzzle

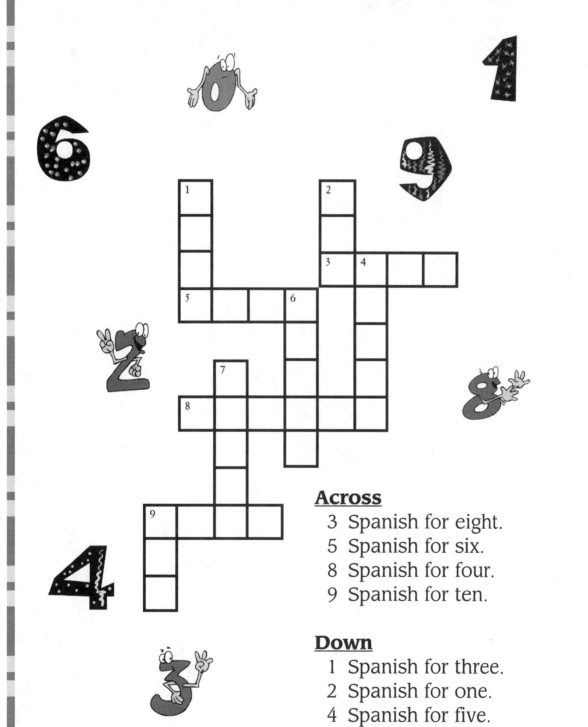

Across

3 Spanish for eight.
5 Spanish for six.
8 Spanish for four.
9 Spanish for ten.

Down

1 Spanish for three.
2 Spanish for one.
4 Spanish for five.
6 Spanish for seven.
7 Spanish for nine.
9 Spanish for two.

Word Search

```
p  s  o  m  b  r  e  r  o  g  i  h  g  v
s  f  e  v  z  a  p  i  n  a  t  a  r  k
e  q  w  s  e  n  o  r  b  t  y  x  a  c
n  h  b  o  f  m  p  a  m  i  g  o  c  y
o  x  u  r  i  o  m  i  j  m  t  q  i  x
r  f  r  n  e  u  e  y  c  n  o  u  a  n
i  y  r  z  s  s  x  i  a  q  s  z  s  f
t  h  o  x  t  d  i  s  c  k  t  l  n  c
a  a  p  i  a  j  c  g  t  d  a  i  i  h
v  t  a  c  o  f  o  c  u  p  d  y  k  a
r  j  o  c  x  g  s  f  s  z  a  h  m  r
v  o  l  c  a  n  o  i  e  w  m  w  q  r
d  m  g  v  s  w  n  b  u  r  r  i  t  o
e  n  c  h  i  l  a  d  a  y  p  h  q  e
```

amigo	enchilada	senorita
burrito	fiesta	sombrero
burro	gracias	taco
cactus	Mexico	tostada
charro	pinata	volcano
	senor	

North America Review Map

See how many countries you can identify. Write their names on the map.

© 2003 Geography Matters

ad maiorem Dei gloriam!

159

South America

South America is the fourth largest continent with the fifth largest population. The west coast is dominated by the mighty Andes Mountains, which run the length of the continent from Columbia to Tierra del Fuego. Both the highest point and lowest point in South American are in Argentina. Cerro Acancagua is 22,831 ft. high and the lowest point, Salinas Chicas, is 138 ft. below sea level. South America is home to the Amazon River, which is surrounded by the world's largest rain forest.

South America

Color each country you study.

Caribbean Sea

Venezuela

Guyana

Suriname

French Guiana

Colombia

Ecuador

Peru

Brazil

Bolivia

Atlantic Ocean

Paraguay

Chile

Uruguay

Argentina

Falkland Islands

© 2003 Geography Matters   ad maiorem Dei gloriam!

South America

General Reference Books

- [ ] Pages 54-61 *Our Father's World*
- [ ] *South America* (A True Book) by David Petersen
- [ ] *Magellan and the Exploration of South America* by Colin Hynson
- [ ] *South American Animals* by Caroline Arnold

Literature

- [ ] *Kitten in the Well* (Rod & Staff)
- [ ] *Capyboppy* by Bill Peet
- [ ] Chapters 4 & 19 – *Heaven's Heroes*
- [ ] Chapter 11 - *Missionary Stories with the Millers*

Science

- [ ] Anteater
  - *Anteaters, Sloths, and Armadillos* by Ann O. Squire
  - Page 26 – *Special Wonders of the Wild Kingdom*
  - http://encarta.msn.com/find/MediaMax.asp?pg=3&ti=761572531&idx=461516528

  Notebook Suggestions:
    1. What do anteaters eat?
    2. How do anteaters use their claws?
    3. Describe the anteater's tongue.
    4. Where do anteaters sleep?

- [ ] Chinchilla
  - http://encarta.msn.com/find/Concise.asp?z=1&pg=2&ti=761566491
  - http://encarta.msn.com/find/MediaMax.asp?pg=3&ti=761563071&idx=461546808 (membership required.

  Notebook Suggestions:
    1. Describe chinchilla fur.
    2. Where do chinchillas live?
    3. How do chinchillas get water?
  - Many pet stores have chinchillas. Go on a field trip and let the children see (and maybe even pet) a chinchilla.

- [ ] Giant Armadillo
  - *Anteaters, Sloths, and Armadillos* by Ann O. Squire
  - http://encarta.msn.com/find/MediaMax.asp?pg=3&ti=761558839&idx=461534280

Notebook suggestions:

1. Where are giant armadillos found?
2. What does the word "armadillo" mean?
3. What are scutes?
4. What do armadillos use their claws for?

❑ Jaguar

- http://encarta.msn.com/find/Concise.asp?z=1&pg=2&ti=761554717

Notebook Suggestions:

1. Where do jaguars live?
2. Why are jaguars hunted?
3. Can jaguars swim?

❑ Sloth

- Pages 60 - *Special Wonders of the Wild Kingdom*
- *Tropical Forest Animals* by Elaine Landau
- *Anteaters, Sloths, and Armadillos* by Ann O. Squire
- www.northcoast.com/~ccar/sloth.htm
- www.geocities.com/Hollywood/Set/1478/sloth.html
- http://encarta.msn.com/find/MediaMax.asp?pg=3&ti=761562271&idx=461538552

Notebook Suggestions:

1. How does a sloth spend most of it's life?
2. How long is a sloth?
3. Why is this animal called a sloth?

❑ Birds

- *Usborne First Nature: Birds*
- *How Does a Bird Fly?* – Usborne Starting Point Science
- *How Do Birds Find Their Way?* by Roma Gans
- *When Birds Change Their Feathers* by Roma Gans
- *A Nest Full of Eggs* by Priscilla Belz Jenkins
- *The Bird Book* by Laura Storms
- Lesson 18 – *Considering God's Creation*
- Build a birdbath, birdhouse, or feeding tray.

Notebook Suggestions:

1. What is a bird's mouth called?
2. What is a bird's chest called?
3. What is a bird's head called?
4. What are the long feathers at the end of the wing called?
5. Name three birds that cannot fly.
6. Name a bird that can fly backwards.
7. What bird chisels wood with its beak?
8. What is the name of your state bird?
9. Name three or four birds found at your house.

Bible

MAZE CRAZE

Start

MAZE CRAZE

Help the bird
into the house.

Start
Here →

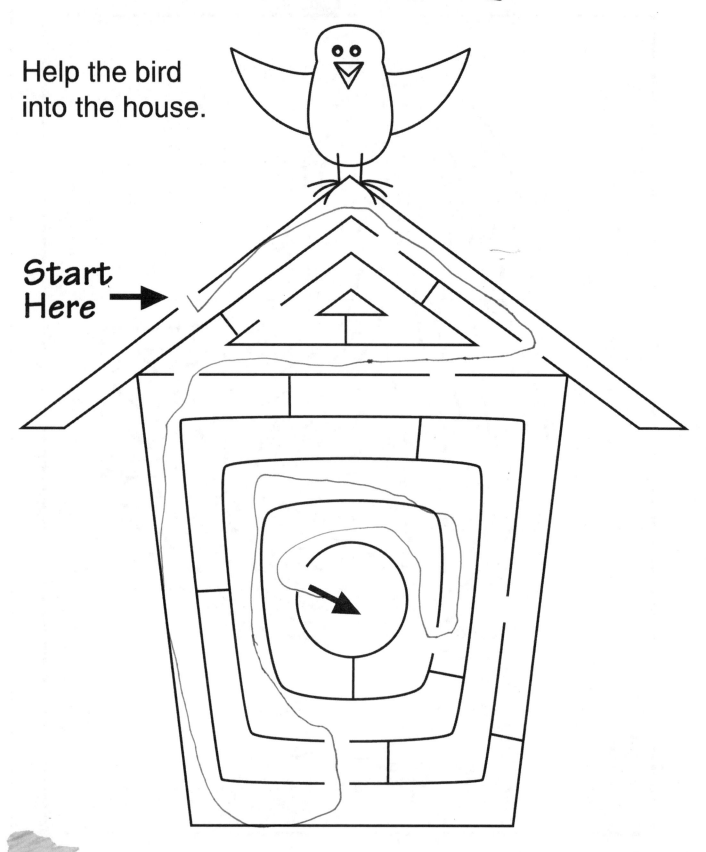

Word Search

```
f o r e s t n s j f v
z p a v a d t t f w e
p f i u b r a z i l n
e m n v m z b n a c e
r d t i x y m h r j z
u k s o u t h m a s u
v a m e r i c a c x e
p o w z j q h g y z l
y y g z k c a v t f a
p a r g e n t i n a m
v o e n f x z p w r l
```

America   Peru
Argentina   rain
Brazil   South
forest   Venezuela

Brazil

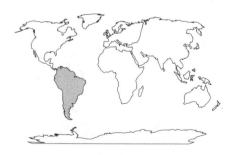

Population_____

Capital City_____

Religion_____

Type of Government_____

Currency_____

Language_____

What are the people called?_____

168

Brazil

Brazil is the fifth largest country in the world and makes up nearly half of the total area of South America. Brazil borders every country in South America except Chile and Ecuador. The Portuguese spoken in Brazil has been greatly influenced by the African and native South American dialects present during the colonization of Brazil. Brazilians take pride in living in an independent country. September 7, 1822, is the historic date when Prince Dom Pedro of Portugal, who became Brazil's first emperor, proclaimed "Ipirange" (Independence) – "Brazilians, our motto will be Independence or Death!"

Since most of Brazil lies in the tropics, the weather is generally hot and humid, but its mountains are cool and sometimes even have frost or snow. During June, July, and August, which are the winter months in South America, as much as 100 inches of rain may fall in the Amazon River basin. The mighty Amazon River is fed by thousands of waterways, which seep through the tropical growth of the world's largest rain forest, covering an area of over 815 million acres. Broad and muddy, the Amazon flows eastward to the Atlantic Ocean. Tropical fruit, like bananas, melons, pineapples, mangoes, and oranges, are grown throughout Brazil. Almost every Brazilian meal includes fruit.

Soccer is the national sport of Brazil.

General Reference Books
- ❑ *Brazil* (A True Book) by Ann Heinrichs
- ❑ *Count Your Way Through Brazil* by Jim Haskins
- ❑ *A Family in Brazil*
- ❑ Pages 14 & 15, *Children Just Like Me* by Barnabas & Anabel Kindersley

Vocabulary

canopy	camouflage	ecosystem
emergent layer	understory	forest floor

Science
- ❑ Rain Forest
  - *Wonders of the Rain Forest* by Janet Craig
  - *Rain Forests* by Anna O'Mara
  - *One Day in the Tropical Rain Forest* by Jean Craighead George
  - http://encarta.msn.com/find/MediaList.asp?pg=6&mod=2&ti=761552810

Music/Art/Projects

1. Color the flag of Brazil.

    Brazil's flag displays its motto, "Order and Progress". Green represents the rainforest. Yellow represents the country's mineral resources. The blue of the globe and the white of the stars represent Brazil's ties to Portugal. The stars represent the states.

169

2.   Color or make a map of Brazil.

3.   Play soccer.

4.   Play ferol bola.

To make your own ferol bola game, all you need are two wooden paddles (the actual ones are slightly larger than ping pong paddles), and a hard rubber or plastic ball. You can substitute the ball with a badminton bird. Mark off a "court" in the sand, yard, sidewalk, or driveway, with a dividing line down the center. The object of the game is to hit the ball back and forth over the line without letting it hit the ground or go out of bounds.

5.   Make your own miniature rain forest using an aquarium, fishbowl, or wide-mouthed glass jar.

Cover the bottom with small stones, sand or charcoal to absorb the water; then add a thick layer of potting soil or dirt. Gently poke in small plants, ferns, and mosses.

Water your plants only when necessary. You may leave the top open or cover it with a piece of plastic wrap or glass, removing the cover every so often if too much moisture accumulates. Pinching back the plants when they have grown too tall will make them fuller and more beautiful.

6.   Make Feijoada.

2 cups dry black beans	1 teaspoon salt
1/2 teaspoon black pepper	2 cloves garlic, minced
4 cups water	2 ounces salt pork
2 1/2 cups canned tomatoes	1 onion, chopped
12 ounces chorizo sausage	

Directions: Soak black beans overnight; drain. Combine soaked beans with the rest of the ingredients except sausage. Bring to a gentle boil. Simmer 1 1/2 hours, adding more water if necessary. Add sausage and cook uncovered until the liquid has thickened and sausage is cooked. Serve over rice with cold orange slices on the side.

7.   Make brigaderiros.

2 tablespoons margarine	1 can sweetened condensed milk
2 tablespoons cocoa	chocolate sprinkles

Directions: Mix the margarine, milk, and cocoa together. Cook over low heat, stirring continuously until thick. Remove from heat and cool completely. Grease your hands with margarine and roll the chocolate into small balls. Roll the ball in chocolate sprinkles.

Internet Resources

❑   http://encarta.msn.com/find/MediaList.asp?pg=6&mod=2&ti=761554342

Flag of Brazil

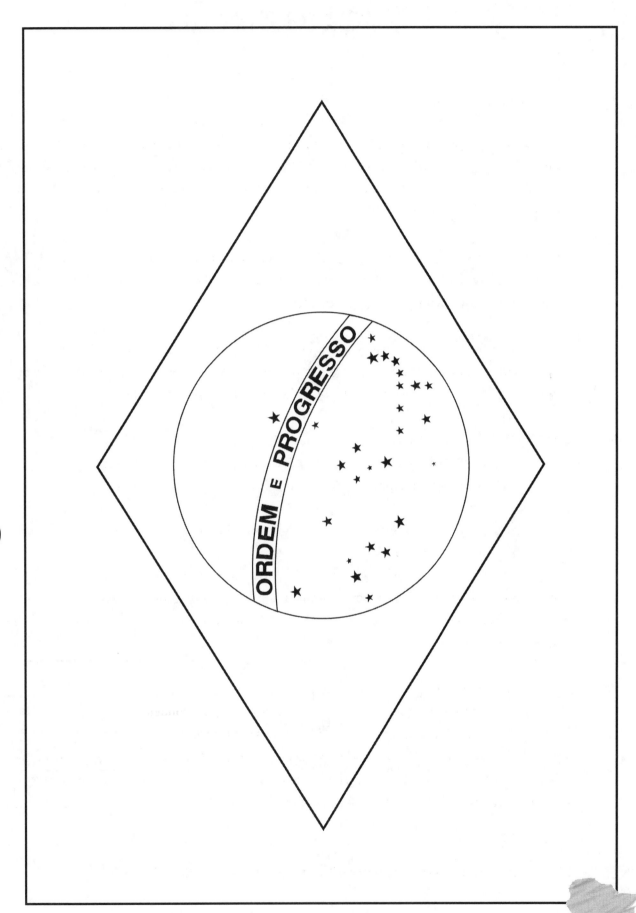

ORDEM E PROGRESSO

Venezuela

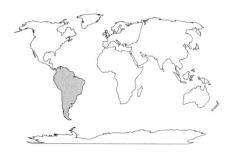

Population_____

Capital City_____

Religion_____

Type of Government_____

Currency_____

Language_____

What are the people called?_____

Venezuela

Venezuela ("Little Venice") was given its name because the vast, shallow inlet of Lake Maracaibo reminded early European explorers of the lagoons around Venice, Italy. Venezuela is located entirely within the tropics; however due to its high altitude some mountain peaks remain permanently covered with snow. The Orinoco River dominates Venezuela. It flows out of the Andes Mountains and Guiana Highlands through the grassy plains of the Llanos to form a wide, swampy delta on the Caribbean coast. The highest waterfall in the world, Angel Falls, is in Venezuela. American pilot, Jimmy Angel discovered the falls in 1935. Angel Falls, at 3,212 feet, is fifteen times higher than Niagara Falls.

Venezuela is rich in mineral resources, especially petroleum reserves. There are significant deposits of iron, bauxite, coal, copper, diamonds, nickel, manganese, and gold.

General Reference Books
- ❑ *Venezuela* (A True Book) by Ann Heinrichs
- ❑ *Take a trip to Venezuela* by Keith Lye

Literature
- ❑ Venezuelan Poem

One Little Elephant
One little elephant,
Out for a run,
Climbed up a spider's web
Just for fun.
He tiptoed across,
He did a little dance,
And then he called down
For some more ele-phants.

Two little elephants,
Out for a run...

Three little elephants...

Vocabulary
iron ore

Music/Art/Projects

1. Color the flag of Venezuela.

    Blue represents Venezuela's independence from Spain. The seven stars represent the seven provinces that supported independence. Red symbolizes courage.

2. Color or make a map of Venezuela.

3. Do a ceramics project.

4. Make Quesillo

1/4 cup sugar	1/8 cup water
4 eggs	1 can evaporated milk
1 can sweetened condensed milk	1 teaspoon vanilla

    Directions: In an ovenproof pan, boil the sugar and water, tilting the pan to coat. Do not brown. Mix the remaining ingredients in a blender. Pour the mixture into the caramelled pan. Refrigerate, then bake for 1 hour at 350° or until a knife inserted into the quesillo comes out clean.

Internet Resources

❑ http://encarta.msn.com/find/MediaList.asp?pg=6&mod=2&ti=761560608

Flag of Venezuela

Peru

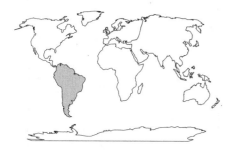

176

Population_____

Capital City_____

Religion_____

Type of Government_____

Currency_____

Language_____

What are the people called?_____

Peru

The Andes Mountains run the length of western Peru. Lake Titicaca, the largest lake in South America, lies on the border between Peru and her neighbor to the south, Bolivia. Lake Titicaca is actually made up of two smaller lakes, Chucito and Uinamarca, which are connected by a narrow strait. Tall, tough reeds called totora grow on the shores of Lake Titicaca. They are used to make houses and fishing boats. The fishing boats are called balsas.

General Reference Books
- ❑ *A Family in Peru* by Jetty St. John
- ❑ *Take a Trip to Peru* by Keith Lye
- ❑ *My Amazon River Day* by Kris Nesbitt

Literature
- ❑ Chapter 14 – *Missionary Stories with the Millers*

Vocabulary

rickshaw   totora   balsas

Music/Art/Projects

1. Color or make the flag of Peru.
    Red and white are the colors chosen by San Martin, "El Liberador" (the Liberator). The colors also recall those of the Incas, who ruled much of Peru until European colonization.

2. Color or make a map of Peru.

3. Make Peruvian Caramel Sauce with Fruit.

12 oz. evaporated milk	2 cups milk
1/2 teaspoon baking soda	1 1/2 packed brown sugar
1/4 cup water	fruit

    Heat evaporated milk, milk, and baking soda to boiling; remove from heat. Heat brown sugar and water in Dutch oven over low heat, stirring constantly, until sugar is dissolved. Add milk mixture. Cook uncovered over medium-low heat, stirring frequently, until mixture is very thick and golden brown, about 1 hour. Pour into serving bowl. Cover and refrigerate at least 4 hours. Serve with fruit.

Internet Resources
- ❑ http://encarta.msn.com/find/MediaList.asp?pg=6&mod=2&ti=761570790

Bible
- ❑ Chapter 14 - *Missionary Stories with the Millers*
  - • I Thessalonians 5:18

Flag of Peru

Argentina

Argentina gets its name from the Latin word argentums, meaning "silver". Unlike many South American countries, there are not great differences between the very rich and the very poor. Most Argentines consider themselves to be middle class. There are; however, unfriendly differences between city dwellers and country people. Iguazu Falls, the world's widest waterfall, is in Argentina. Iguazu Falls is over two miles wide and gets its name from the Indian words for "great waters".

General Reference Books

- ❏ *Argentina* by Michael Burgan
- ❏ Pages 12 & 13, *Children Just Like Me* by Barnabas & Anabel Kindersley

Literature

- ❏ *On the Pampas* by Maria Cristina Brusca
- ❏ *My Mama's Little Ranch on the Pampas* by Maria Cristina Brusca
- ❏ Argentine Poem

Kingdom of Reverse

I was told that in the Kingdom of Reverse
There are swimming cats and flying fish,
The cats don't cry out loud
And call out "yes" learning English in their test.
Let's see what goes on in the Kingdom of Reverse.

I was told that in the Kingdom of Reverse
No one dances on their feet.
One rat is the night watchman,
The other a judge;
And that two plus two is three.
Let's see what goes on in the Kingdom of Reverse.

I was told that in the Kingdom of Reverse
A bear fits in a nut,
The kids are grown-ups, and little girls wear makeup.
And a week feels like a year.
Let's see what goes on in the Kingdom of Reverse.

I was told that in the Kingdom of Reverse
There is a cute Pekinese
That falls up
And gets lost flying down.
Let's see what goes on in the Kingdom of Reverse.

Argentina

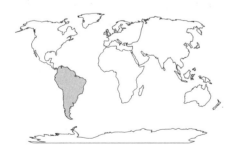

Population_____

Capital City_____

Religion_____

Type of Government_____

Currency_____

Language_____

What are the people called?_____

I was told that in the Kingdom of Reverse

A lady was named Andrew,

That she has 1500 chimpanzees

And if you look, there is not a chimp to be seen.

Let's see what goes on in the Kingdom of Reverse.

I was told that in the Kingdom of Reverse

That a spider and a centipede

Talk of love and play the fiddle

Riding a chessboard.

Let's see what goes on in the Kingdom of Reverse.

Vocabulary

pampas	bola	overseer
ranch	gaucho	

Music/Art/Projects

1.　Color the flag of Argentina.

Blue and white were formed into a flag by Manuel Belgrano, the leader of the revolution in which Argentina achieved its independence from Spain in 1816. Blue and white recall the sky when the first uprising for independence was staged. The Sun of May was added in 1818 to create a flag for state use.

2.　Color or make a map of Argentina.

3.　Make Dulce de Leche (Sweet Milk Dessert)

　　　2 cans sweetened condensed milk

　　　ice cream or butter cookies

Directions: Shake cans of milk. Place unopened cans in a saucepan and cover completely with water. Boil the cans for an hour and a half, making sure the cans are always covered with water. Do not open cans until they are completely cool. Pour the butterscotch-type sauce over ice cream or spread between two butter cookies.

Internet Resources

❑　http://encarta.msn.com/find/MediaList.asp?pg=6&mod=2&ti=761556250

Bible

❑　*On the Pampas*

　• Matthew 7:11

　• Proverbs 12:10

181

Flag of Argentina

South America Review Map

See how many countries you can identify. Write their names on the map.

© 2003 Geography Matters

ad maiorem Dei gloriam!

Africa

Africa is the second largest continent and has the second largest population. Africa boasts the world's largest desert, the Sahara, and the world's longest river, the Nile. Africa is a land of rain forests, grassy plains and deserts. The highest point in Africa is at Kilamanjaro, Tasmania (19,340 ft) and its lowest point is Lac Assal, Dijbouti located 515 ft. below sea level. Africa is bordered by the Indian Ocean on the east and the Atlantic Ocean on the west.

Map of Africa

Color each country you study.

© 2003 Geography Matters

185

Africa

Africa is the second-largest continent, covering about 20% of the world's land area. Most of Africa is a plateau, surrounded by narrow coastal plains. Kilimanjaro, an inactive volcano in Tanzania, is the highest peak. The Nile, Congo, and Niger are the major rivers. The continent has two main people groups. North of the Sahara are the Arabs and Berbers who speak Arabic and practice Islam. South of the Sahara are the black Africans that are divided into over 1,000 ethnic groups. Africa contains many of the world's poorest countries.

General Reference Books
❏ *The Asante of West Africa* by Jamie Hetfield
❏ Pages 85-92 *Our Father's World*

Literature
❏ *Africa Calling* by Daniel Alderman
❏ *Ashanti to Zulu* by Margaret Musgrove
❏ *Uncommon Traveler: Mary Kingsley in Africa* by Don Brown
❏ *Where Are You Going, Manyoni?* by Catherine Stock
❏ *Zzng! Zzng! Zzng!* by Phyllis Gershator
❏ *A Country Far Away* by Nigel Gray
❏ *Koi and the Kola Nuts* by Verna Aardema
❏ *Off to the Sweet Shores of Africa* by Uzo Unobagha
❏ Chapters 1, 4, & 7 – *Missionary Stories with the Millers*
❏ Chapters 1, 6, & 10 – *Heaven's Heroes*

Science
❏ *African Animals* by Caroline Arnold
❏ *Jungle Jack Hanna's Safari Adventure* by Jack Hanna and Rick A. Prebeg
❏ *On Safari* by Tessa Paul

❏ Giraffe
• *The Giraffe: A Living Tower* by Christine & Michel Denis-Huot
• Page 30 – *Special Wonders of the Wild Kingdom*
• http://encarta.msn.com/find/MediaMax.asp?pg=3&ti=761561060&idx=461562797 (membership required)
Notebook Suggestions:
1. How tall are giraffes?
2. What do giraffes eat?
3. How do giraffes bend down to get a drink?

❏ Apes and Monkeys
• *Monkeys and Apes* (A New True Book) by Kathryn Wentzel Lumley
• http://encarta.msn.com/find/MediaList.asp?pg=6&mod=2&ti=761556424
• http://encarta.msn.com/find/MediaList.asp?pg=6&mod=2&ti=761569669

Notebook Suggestions:

1. In what ways are monkeys and apes different?

2. How are monkeys and apes similar?

❑ Chimpanzees

• *Chimpanzees* (A True Book) by Patricia A. Fink Martin

Notebook Suggestions:

1. Is a chimpanzee a monkey or an ape?

2. Where are chimpanzees found?

3. What do chimpanzees eat?

4. What can chimpanzees do that most other animals cannot do?

❑ Gorillas

• *Gorillas* (A True Book) by Patricia A. Fink Martin

• *Gorillas: Gentle Giants of the Forest* by Joyce Milton

• Page 32 – *Special Wonders of the Wild Kingdom*

• www.gorilla-haven.org

• http://encarta.msn.com/find/MediaList.asp?pg=3&ti=761566978

Notebook Suggestions:

1. What makes up a gorilla family group?

2. What is this group called?

3. Where are gorillas found?

4. Where do gorillas sleep?

5. What do gorillas eat?

6. How big is a baby gorilla at birth?

❑ Hippopotamus

• Page 36 – *Special Wonders of the Wild Kingdom*

• *Hippos* by Sally M. Walker

• *Hippos* by Miriam Schlein (You may want to skip the first two chapters due to evolutionary content.)

• http://encarta.msn.com/find/MediaList.asp?pg=6&mod=2&ti=761563889

Notebook Suggestions:

1. Where do hippos live?

2. What do hippos eat?

3. What are baby hippos called?

4. Describe a hippo's skin.

5. What do hippos do during the day?

❑ Lions

• Page 42 - *Special Wonders of the Wild Kingdom*

• *Lions* by Cynthia Overbeck

• *Lion* by Caroline Arnold

• http://encarta.msn.com/find/MediaList.asp?pg=6&mod=2&ti=761566718

Notebook Suggestions:

1. Why do male lions have manes?

2. When do lions hunt?

3. What is a group of lions called?

4. What is a baby lion called?

❑ Ostrich

• *Ostriches* (A New True Book) by Emilie U. Lepthien

• http://encarta.msn.com/find/MediaMax.asp?pg=3&ti=761575132&idx=461534626

Notebook Suggestions:

1. How tall is an adult ostrich?

2. How fast can an ostrich run?

3. Where do ostriches live?

4. What do ostriches eat?

5. Why do ostriches stick their heads in the sand?

❑ Rhinoceros

• Page 16 – *Special Wonders of the Wild Kingdom*

• http://encarta.msn.com/find/Concise.asp?z=1&pg=2&ti=761552776 (May want to edit due to evolutionary content.)

• http://encarta.msn.com/find/MediaList.asp?pg=6&mod=2&ti=761552776

Notebook Suggestions:

1. How does a rhinoceros use its front horn?

2. What do rhinoceroses eat?

3. Which sense do rhinoceroses rely on most?

❑ Zebras

• Page 68 - *Special Wonders of the Wild Kingdom*

• http://encarta.msn.com/find/MediaList.asp?pg=6&mod=2&ti=761575094

Notebook Suggestions:

1. Where are zebras found?

2. What do zebras do when they are in danger?

3. What animal is the zebra related to?

4. How do zebras spend most of their time?

Bible

- ❏ Chapter 1 - *Missionary Stories with the Millers*
  - Psalm 91
- ❏ Chapter 4 - *Missionary Stories with the Millers*
  - Exodus 15:26b
  - Matthew 4:23b
- ❏ Chapter 7 - *Missionary Stories with the Millers*
  - Matthew 9:27-31
  - Nehemiah 8:10
- ❏ Lion
  - Proverbs 28:1
  - Psalm 17:12
  - Amos 3:4
  - Nahum 2:11 & 12
  - Proverbs 30:30
  - Daniel 6
- ❏ Apes
  - 2 Chronicles 9:21
- ❏ Ostrich
  - Lamentations 4:3
  - Job 39:13-18

¡FIND THE TWINS!

Which two are exactly alike?

3

6

2

5

1

4

190

MAZE CRAZE

Crossword Puzzle

Across

3 I am like a monkey, but I do not have a tail.

4 I am the largest member of the ape family.

6 I am like an ape, but I have a tail.

9 I am a member of the ape family.

10 I am the second largest land mammal. Only the elephant is bigger.

Down

1 I have black and white stripes.

2 I have a long neck.

5 I have a thick mane.

7 I am a bird that cannot fly.

8 I have a big horn coming out of my forehead.

Word Search

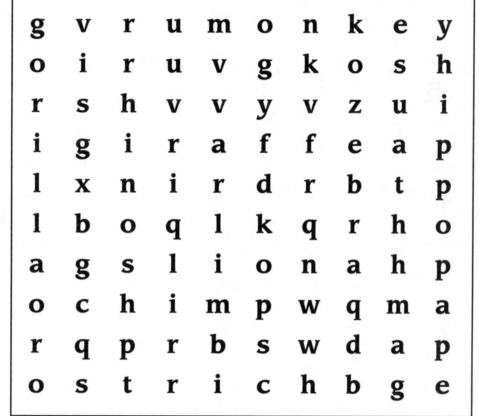

g	v	r	u	m	o	n	k	e	y
o	i	r	u	v	g	k	o	s	h
r	s	h	v	v	y	v	z	u	i
i	g	i	r	a	f	f	e	a	p
l	x	n	i	r	d	r	b	t	p
l	b	o	q	l	k	q	r	h	o
a	g	s	l	i	o	n	a	h	p
o	c	h	i	m	p	w	q	m	a
r	q	p	r	b	s	w	d	a	p
o	s	t	r	i	c	h	b	g	e

ape   lion
chimp   monkey
giraffe   ostrich
gorilla   rhino
hippo   zebra

South Africa

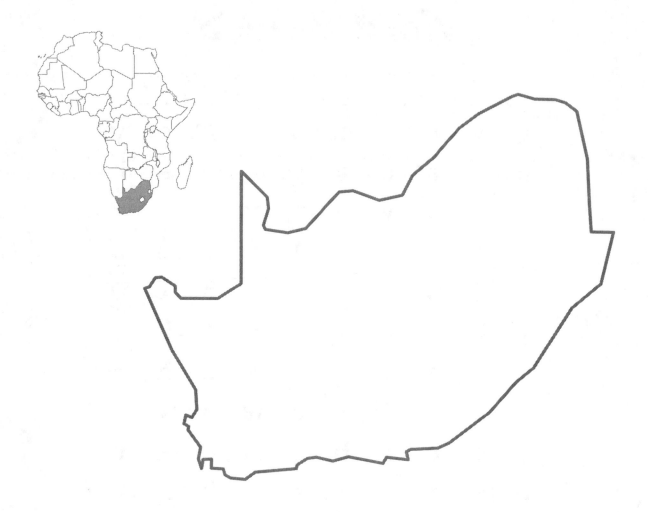

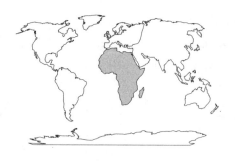

194

Population_____

Capital City_____

Religion_____

Type of Government_____

Currency_____

Language_____

What are the people called?_____

South Africa

South Africa is extremely rich in wildlife. Elephants, rhinoceros, zebras, antelopes, giraffes, and other wildlife are protected in game reserves such as Kruger National Park, Hluhluwe, and Mkuze. There are more than 100 species of snakes native to South Africa. In some places in South Africa, wildlife still roam freely. Wild baboons can sometimes be seen on top of buses. Signs reading "Don't Feed the Baboons" are common.

South Africa is home to the baobab tree. The trunks of the trees are so thick that in some cases their distance around is greater than their height. It also looks as if the baobab tree grows upside down with its roots in the air instead of its branches.

There are three capitals of South Africa. The legislative branch of the government is in Cape Town. The administrative branch is in Pretoria. The judicial branch is located in Bloemfontein. South Africa is a leading producer of diamonds and gold.

People/History
❑ Nelson Mandela
 • Page 217 - *Usborne Book of Famous People*

General Reference Books
❑ *The Zulu of Southern Africa* by Christine Cornell
❑ *South Africa* (A New True Book) by Karen Jacobsen

Literature
❑ Chapter 24 – *Missionary Stories with the Millers*
❑ *Jaffa's Journey* by Hugh Lewin
❑ *Jaffa's Mother* by Hugh Lewin
❑ *Jaffa's Father* by Hugh Lewin

Vocabulary
village   courtyard   climate

Music/Art/Projects
1.   Color the flag of South Africa.
       The flag of South Africa was adopted in 1994. It includes the colors of all the parties that helped to end apartheid. Black, yellow, and green are the colors of the major black parties. Red, white, and blue are from the British and Dutch flags. The "Y" shape represents the union of the two sides into a unified entity.

2.   Color or make a map of South Africa.

Internet Resources
❑ http://encarta.msn.com/find/MediaList.asp?pg=6&mod=2&ti=761557321

Bible
❑ Chapter 24 - *Missionary Stories with the Millers*
 • 2 Corinthians 5:17

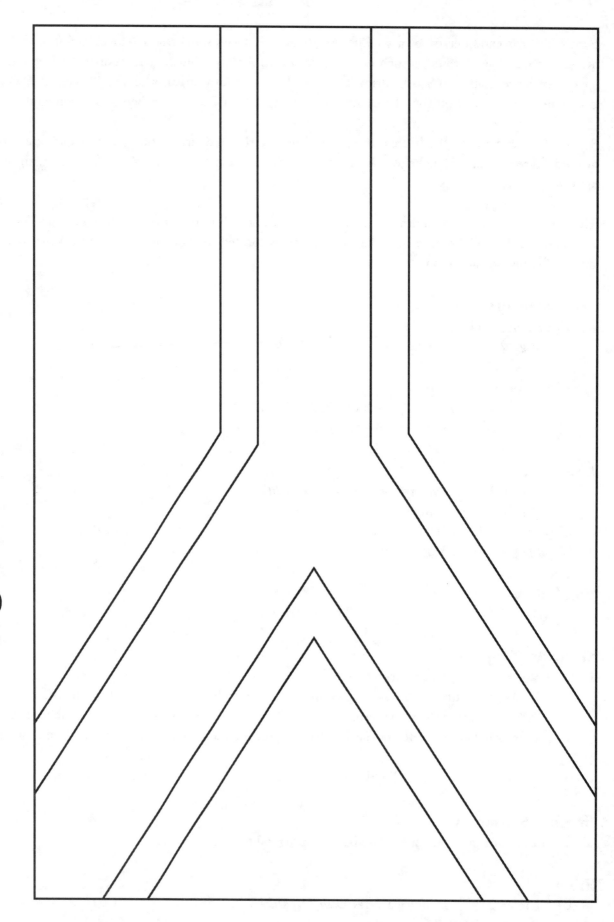

Flag of South Africa

Kenya

Kenya is an agricultural nation. Farming and ranching are Kenya's main occupations. Corn is the most widely grown crop and is the main food of the Kenyan diet. The life-style in Kenya's large cities is modern, but in the remote areas, people live homes made of dried mud and thatch with no modern conveniences. Swahili is the language of Kenya.

General Reference Books
- ❑ *Next Stop Kenya* by Fred Martin
- ❑ *A Family in Kenya*

Literature
- ❑ *Moja Means One: Swahili Counting Book* by Muriel Feelings

Vocabulary

safari   wildlife refuge   sanctuary

Music/Art/Projects

1. Color the flag of Kenya.

    The flag of Kenya has four colors. The top black stripe stands for the Kenyan people, the middle red stripe for their blood and the struggle for independence, and the green stripe for the rich land. The two white stripes represent peace, unity, and the non-African minorities. The Masai war shield in the center symbolizes the defense of freedom.

2. Color or make a map of Kenya.

3. Make Irio

4 cups corn	8 potatoes, mashed
1 cup peas	

    Directions: Bring corn and peas to a boil. Reduce heat and simmer until done. Drain. Mix with mashed potatoes. Add salt to taste.

Internet Resources
- ❑ http://encarta.msn.com/find/MediaList.asp?pg=6&mod=2&ti=761564507

Kenya

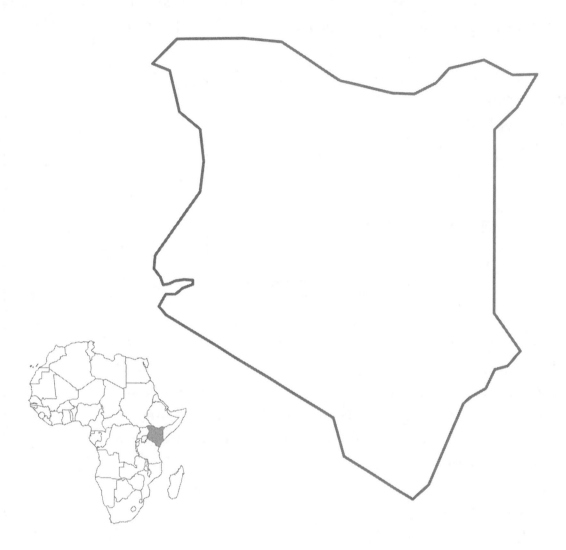

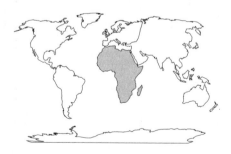

Population_____

Capital City_____

Religion_____

Type of Government_____

Currency_____

Language_____

What are the people called?_____

198

Flag of Kenya

199

Morocco

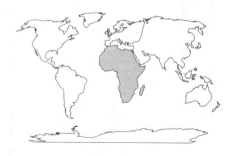

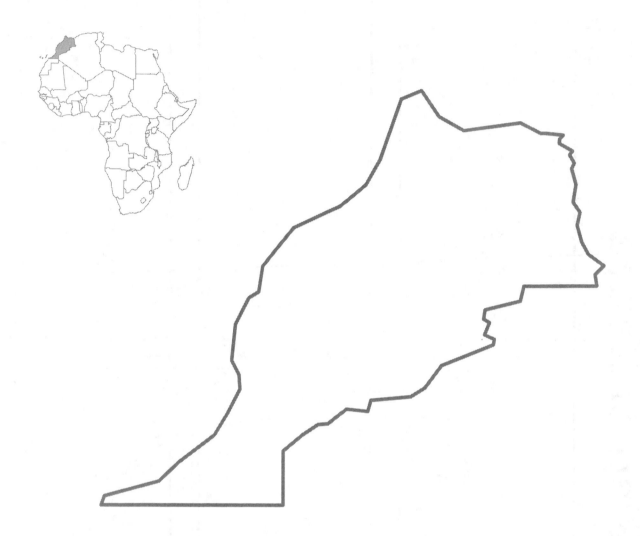

Population_____

Capital City_____

Religion_____

Type of Government_____

Currency_____

Language_____

What are the people called?_____

Morocco

Moroccans are mostly of Arab and Berber descent. Arabic is the official language of Morocco; however, Spanish, French, and English are widely spoken. Islam is the official religion. The vast majority of Moroccans are Sunnite Muslims.

As with many former European colonies in Africa, the economy of Morocco relies heavily on the export of raw materials. Two-thirds of the world's phosphate reserves lie in Morocco. Other minerals include iron ore, lead, zinc, coal, and manganese. An abundant supply of tuna, sardines, and bonito can be found off Morocco's west coast.

General Reference Books
- ❑ *A Family in Morocco* by Judy Stewart
- ❑ *Morocco* by Bob Italia
- ❑ *Take a Trip to Morocco* by Keith Lye
- ❑ *The Children of Morocco* by Jules Hermes
- ❑ Pages 40 & 41, *Children Just Like Me* by Barnabas & Anabel Kindersley

Literature
- ❑ *Ali, Child of the Desert* by Jonathan London

Vocabulary

medina	tile	alleyway
mosque	mint	kiln
desert	Berber	

Music/Art/Projects

1. Color the flag of Morocco.

    Red represents the descendants of the Prophet Muhammad. From the 17th century on, when the Hassani Dynasty ruled Morocco, the flag of the country was plain red. In 1915, during the reign of Mulay Yusuf, the green seal of Solomon was added to the national flag. The seal is an interlaced pentangle, used as a symbol in occult law for centuries.

2. Color or make a map of Morocco.

3. Make Couscous with Chicken.

2 tablespoons olive oil	3 lb chicken, cut up
4 small carrots, sliced	2 medium onions, sliced
2 medium turnips, quartered	2 cloves garlic, minced
2 teaspoons ground coriander	1 1/2 teaspoons salt
1 teaspoon chicken bouillon	1/4 teaspoon red pepper

1/4 teaspoon ground turmeric	1 cup water
3 zucchini, sliced	15 oz. can garbanzo beans
Couscous	

Heat oil in Dutch oven until hot. Cook chicken in oil until brown on all sides, about 15 minutes. Drain fat from Dutch oven. Add carrots, onions, turnips, garlic, coriander, salt, bouillon, ground red pepper, and turmeric. Pour water over vegetables. Heat to boiling; reduce heat. Cover and simmer 30 minutes. Add zucchini to chicken mixture. Cover and cook until chicken is done, about 10 minutes. Add beans; cook 5 minutes.

Prepare couscous. Mound in center of heated platter; arrange chicken and vegetables around Couscous.

1 1/3 cups couscous	3/4 cup raisins
1/2 teaspoon salt	1 cup boiling water
1/2 cup butter	1/2 teaspoon ground turmeric

Mix couscous, raisins, and salt in 2-quart bowl; stir in boiling water. Let stand until all water is absorbed, 2 to 3 minutes. Heat butter in skillet until melted; stir in couscous and turmeric. Cook and stir 4 minutes.

Internet Resources

❑ http://encarta.msn.com/find/MediaList.asp?pg=6&mod=2&ti=761572952

Flag of Morocco

Nigeria

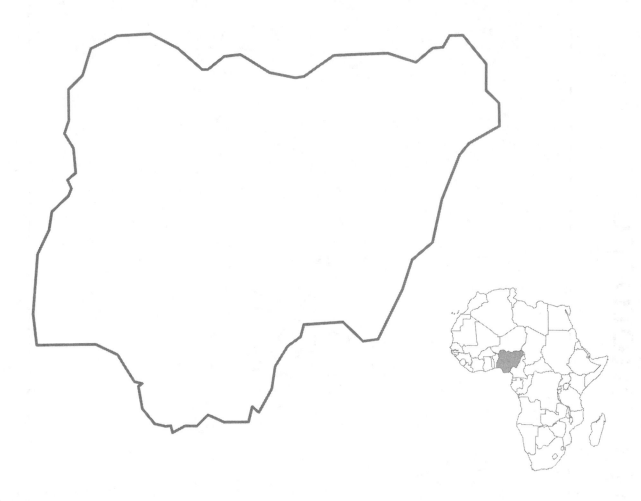

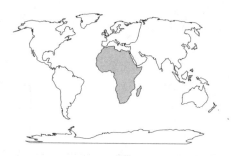

Population_____

Capital City_____

Religion_____

Type of Government_____

Currency_____

Language_____

What are the people called?_____

Nigeria

Nigeria is the size of Texas and New Mexico combined. The Niger River and its tributary, the Benue, form a Y-shape that divides the country into three regions: north, east, and west.

Nigeria is famous for its art. The oldest known African sculptures are clay figures from about 500 B.C. found in central Nigeria. Some of the best woodcarvings in Africa come from Nigeria. Wood carvers called onishona (people who make art), train for years to learn their art.

Yams (the staple of the Nigerian diet), cassava, rice, and plantains are grown in southern Nigeria. Northern Nigeria grows millet, sorghum, peanuts, cotton, and rice. Nigerians love music and it has always been important to the people and their culture. In addition to singing and dancing, drums are popular instruments as well as raft zithers, thumb pianos, and metal gongs. Musicians also play a stringed instrument called the molo, which has three or four strings, and a reed flute called an algaita.

The naming ceremony is a very important family custom. Seven days after the birth of a child, the father gives kola nuts to family and friends as an invitation to join the celebration and praise of the newborn.

General Reference Books
- ❑ *A Family in Nigeria* by Carol Barker
- ❑ *Take a Trip to Nigeria* by Keith Lye
- ❑ *Saying Good-bye* by Ifeoma Onyefulu

Literature
- ❑ A Nigerian fable:

    The chief sent out messengers to announce that hewould give a feast and asked each guest to bring one calabash of palm wine. One man wanted very much to attend, but he had no wine to bring. When his wife suggested that he buy the wine, he said, "What! Spend money so that I can attend a feast that is free?" He thought to himself, "If hundreds of people were to pour their wine into the chief's pot, could just one calabash of water spoil so much wine?

    The day of the feast came. Everyone bathed and dressed in their best clothes and gathered at the house of the chief. There was music and festive dancing. Each man, as he entered the chief's compound, poured the contents of his calabash into a large earthen pot. The man also poured his water there then greeted the chief.

    When all the guests had arrived, the chief ordered his servants to fill everyone's cup with wine. The man was impatient, for there was nothing so refreshing as palm wine. At the chief's signal, all the guests put the cups to their lips and tasted...and tasted again...for what they tasted was not palm wine but water. Each guest had thought that his one calabash of water could not spoil a great pot of palm wine.

Music/Art/Projects

1. Color or make the flag of Nigeria.

   The flag of Nigeria was adopted in 1960. The two green stripes represent the agriculture of Nigeria. The white stripe symbolizes unity and peace.

2. Color or make a map of Nigeria.

3. Play Mancala.

4. Make Banana Fritters.

2 1/2 cups flour	1/2 cup sugar
2 teaspoons cinnamon	2 eggs
1 cup milk	6 bananas, mashed
oil	confectioner's sugar

   Directions: Combine flour, sugar, and cinnamon. Beat in eggs. Gradually add milk and beat until smooth. Stir in bananas. Pour batter by 1/4 cupfuls onto hot griddle. Cook 2-3 minutes each side. Sprinkle with confectioner's sugar before serving.

5. Make a Nigerian drum.

Small clay flowerpot	markers or paint
Paper grocery bag	paper taper

   Directions: Decorate outside of flowerpot with markers or paint. Let dry. Cut a circle from the paper bag 3" bigger than the open end of the flowerpot. Dampen circle and lay it over open end of flower pot. Pull the bag tight and tape it in place. Wrap the tape around several times. Allow to dry and enjoy the beautiful music.

6. Make African beads.

4 cups flour	2 cups salt
2 cups water	large bowl
baking tray	toothpicks
acrylic paints	shellac
string	

   Directions: Mix together flour and salt. Gradually add water and knead dough until thoroughly mixed and smooth. Form small beads out of the dough. Fifteen to twenty beads will be needed for each neck lace. Push a toothpick through the center of each bead to make a hole. Bake 15-20 minutes in a 350° oven. Cool completely. Decorate beads with paint and let dry. Shellac beads and allow to dry over night. Strings beads.

Internet Resources

❑ http://encarta.msn.com/find/MediaList.asp?pg=6&mod=2&ti=761557915

Flag of Nigeria

Egypt

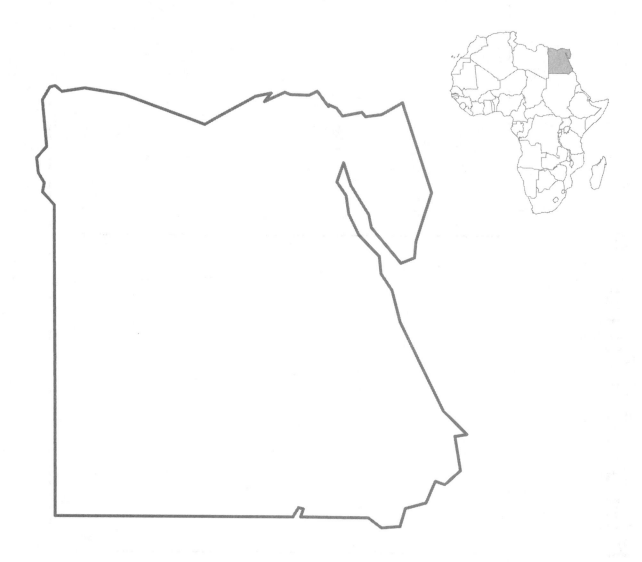

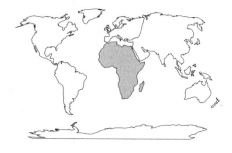

208

Population_____

Capital City_____

Religion_____

Type of Government_____

Currency_____

Language_____

What are the people called?_____

Egypt

Little rain falls in Egypt and most of the country is covered by the Sahara Desert. Ninety-nine percent of Egyptians live along the Nile River Valley and the Suez Canal. Cairo, the capital of Egypt, is the largest city in Africa.

Ancient Egypt is best known for the great pyramids built as tombs for its pharaohs. The most famous pyramids are in Giza. The Great Sphinx of Giza is the largest statue to survive from ancient times. It has the head of a man and the body of a lion. Historians believe the head was built to look like king Kharfe, and that the lion's body stood for the king's strength. Unlike the pyramids, the Great Sphinx was not built for burial. The pyramids were built over a span of 1000 years. The Great Pyramid of Khufu is one of the seven wonders of the ancient world.

Papyrus was the earliest form of paper. Egyptians made the paper using the papyrus reed that grew along the banks of the Nile River.

People/History

❑ Cleopatra
  • *Cleopatra* by Diane Stanley

❑ King Tutankhamen
  • *Tutankhamen* by Robert Green
  • *Tut's Mummy: Lost and Found* by Judy Donnelly

General Reference Books

❑ *A Family in Egypt*
❑ *Take a Trip to Egypt* by Keith Lye
❑ *The Children of Egypt* by Matti A. Pitkanen
❑ Page 36, *Children Just Like Me* by Barnabas & Anabel Kindersley

Literature

❑ *Bill and Pete Go Down the Nile* by Tomie dePaola
❑ *The Day of Ahmed's Secret* by Florence Parry Heide & Judith Heide Gilliland
❑ Chapters 2 & 9 - *Missionary Stories with the Millers*

Vocabulary

pyramid	sphinx	mummy
pharaoh	sarcophagus	market
delta		

Music/Art/Projects

1. *Pyramids! 50 Hands-on Activities to Experience Ancient Egypt* by Avery Hart and Paul Mantell

2. Color the flag of Egypt.
    Red, white, and black are Pan-Arab colors. The eagle in the middle of Egypt's flag is a symbol of Saladin, a Muslim warrior who lived during the 1100's.

3. Color or make a map of Egypt.

4. Make the Rosetta Stone.
    Directions: Roll soft clay into clean meat trays. The children can carve their own hieroglyphs with Popsicle sticks, toothpicks, or the end of a paintbrush. Follow manufacturers directions for drying clay.

5. If you read *The Day of Ahmed's Secret* with a young child, have the child practice writing his name.

Internet Resources

❑ http://encarta.msn.com/find/MediaList.asp?pg=6&mod=2&ti=761557408

Bible

❑ The story of Joseph
   • Genesis 37
   • Genesis 39 – 47
   • Genesis 50:15-26
❑ The story of Moses
   • Exodus 2-4
❑ The plagues of Egypt
   • Exodus 5-11
❑ Passover – the Children of Israel protected
   • Exodus 12:1-30
❑ Children of Israel delivered from slavery.
   • Exodus 12:31-51
❑ Mary and Joseph flee to Egypt
   • Matthew 2:13-15
❑ Chapter 2 - *Missionary Stories with the Millers*
   • Exodus 20:15
   • Acts 16:30-31
❑ Chapter 9 - *Missionary Stories with the Millers*
   • Mark 10:13-17

Flag of Egypt

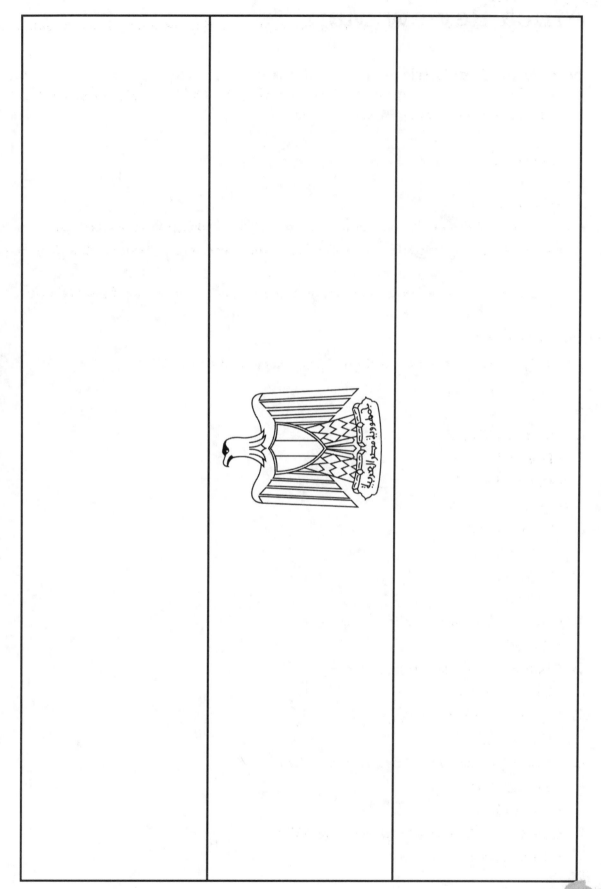

Africa Review Map

See how many countries you can identify. Write their names on the map.

© 2003 Geography Matters

 **Notes:**

Oceania

Oceania includes the countries of Australia and New Zealand and the island groups known as Micronesia, Melanesia, and Polynesia. Oceania stretches from the Tropic of Cancer in the north to New Zealand in the south. The highest point in Oceania is Mt. Wilhelm in Papua New Guinea (14,793 ft.) and its lowest point is at Lake Eyre in South Australia where the elevation is measured 52 ft. below sea level.

Map of Oceania

Color each country you study.

Papua
New Guinea

Australia

New
Zealand

© 2003 Geography Matters

ad maiorem Dei gloriam!

215

Australia

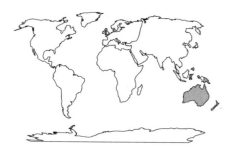

216

Population_____

Capital City_____

Religion_____

Type of Government_____

Currency_____

Language_____

What are the people called?_____

Australia

Originally a group of British colonies, the Commonwealth of Australia still has the British monarch as its head of state. However, there is widespread support for the country becoming a republic. Although cultural ties with Britain are still strong, Australia now sees itself as a pacific nation.

The majority of Australians live in the big cities of the eastern, southern, and southwestern coasts, where the temperature is pleasant and the ground fertile. Cattle and sheep ranches or mines occupy the interior or 'outback'. Cattle and sheep are traditional strengths of the Australian economy. Ranches, known as 'stations', may cover over 6,200 square miles and light aircraft may be needed to cross a single farm.

People/History

- ❑ James Cook
  - Pages 128 & 129 – *Usborne Book of Famous Lives*
  - *A World Explorer: James Cook* by Adele deLeeuw
  - *Captain Cook: Pacific Explorer* by Ronald Syrne

- ❑ Robert Burke & William Wills
  - Pages 130 & 131 – *Usborne Book of Famous Lives*

General Reference Books

- ❑ *Next Stop Australia* by Fred Martin
- ❑ *Passport to Australia* by Susan Pepper
- ☒ *A Family in Australia* by Emily Gunner and Shirley McConky
- ☒ *An Aboriginal Family* by Rollo Browne
- ❑ *Australia* by Mary Berendes
- ☒ Pages 76 & 77, *Children Just Like Me* by Barnabas & Anabel Kindersley
- ☒ *Cooking the Australian Way* by Elizabeth Geramane

Literature

- ☒ *The Pumpkin Runner* by Marsha Arnold
- ☒ *Koala Lou* by Mem Fox
- ☒ *Wombat Stew* by Marcia Vaughn
- ☒ *The Very Boastful Kangaroo* by Bernard Most

Science

- ❑ Coral Reef
  - *Coral Reef* by Barbara Taylor
  - *Down Under Down Under* by Ann McGovern
  - http://encarta.msn.com/find/MediaMax.asp?pg=3&ti=761575831&idx=461543601

❑ Kangaroos

- *Cycle of a Kangaroo* by Angela Royston
- *Young Kangaroo* by Margaret Wise Brown
- http://encarta.msn.com/find/MediaMax.asp?pg=3&ti=761574683 (membership required)

Notebook Suggestions:

1. What does a kangaroo use its tail for?
2. Where do kangaroos live?
3. What is a group of kangaroos called?
4. What is a baby kangaroo called?
5. What is a marsupial?

❑ Wombats

- *Wombats* by Barbara Triggs
- Page 66 – *Special Wonders of the Wild Kingdom*
- http://encarta.msn.com/find/MediaList.asp?pg=3&ti=761566101

Notebook Suggestions:

1. What do wombats use their claws for?
2. How much do adult wombats weigh?
3. When do wombats eat?
4. What do wombats eat?
5. How long can a wombat live?

❑ Koalas

- Page 38 – *Special Wonders of the Wild Kingdom*
- *Koalas* by Denise Burt
- http://encarta.msn.com/find/MediaList.asp?pg=3&ti=761554280

Notebook Suggestions:

1. What do koalas eat?
2. When do koalas climb down a tree?
3. How do koalas defend themselves?
4. What do koalas and kangaroos have in common?
5. How long does a baby koala stay in its mother's pouch?
6. How many hours a day do koalas sleep?

❑ Crocodile

- *Crocodiles!* by Irene Trimble
- *I Wonder Why Crocodiles Float Like Logs* by Annabelle Donati
- http://encarta.msn.com/find/MediaList.asp?pg=3&ti=761578937

Notebook Suggestions:

1. Why do crocodiles lie with their mouth open?
2. How do crocodiles kill their prey?

❑ Tasmanian Devil

- www.tased.edu.au/tot/fauna/devil
- http://encarta.msn.com/find/Concise.asp?ti=065E9000

Notebook Suggestions:

1. How did the Tasmanian devil get its name?
2. What are Tasmanian devils known for?
3. What does the Tasmanian devil eat?
4. How long is a Tasmanian devil?

❑ Duck-billed Platypus

- Page 46 – *Special Wonders of the Wild Kingdom*
- http://encarta.msn.com/find/Concise.asp?z=1&pg=2&ti=761563670

Notebook Suggestions:

1. Describe a platypus.
2. What does a platypus use its claws for?
3. How does a mother platypus hold her babies?
4. What does the platypus eat?

Vocabulary

aborigine	snorkel	shark
coral	outback	marsupial

Music/Art/Projects

1. Color the flag of Australia.

    The Australian flag is red, white, and blue. The British Union flag is at the top representing Australia's ties with Great Britain. The five stars on the right represent the Southern Cross constellation. The larger star on the left is known as the Commonwealth Star. Five of the six stars have seven points. Each point stands for Australia's six states and one territory.

2. Color or make a map of Australia.

3.   Make Damper Bread.

1 package active dry yeast	1/4 cup warm water
2 tablespoons sugar	3 cups all-purpose flour
1 tablespoon baking powder	3/4 teaspoon salt
1/4 cup shortening	1 cup buttermilk

Directions: Dissolve yeast in warm water; stir in sugar. Mix flour, baking powder, and salt in large bowl; cut in shortening until mixture resembles fine crumbs. Stir in yeast mixture and buttermilk. Turn dough onto lightly floured surface; knead gently until smooth, about one minute. Cover; let rise 10 minutes. Shape dough into a round loaf, 6-7 inches in diameter. Place on a greased cookie sheet or pizza pan. Cover; let rise in warm place 30 minutes. Heat oven to 375 ° F. Cut an X about 1/2 inch deep in top of bread. Bake until golden brown, about 35 minutes.

Internet Resources

❏ http://encarta.msn.com/find/MediaList.asp?pg=6&mod=2&ti=761568792

Bible

❏ *The Very Boastful Kangaroo*
- Proverbs 25:14
- Proverbs 27:1
- Proverbs 13:10
- James 4:16

Flag of Australia

FIND THE TWINS

Which two are exactly alike?

3

2

1

6

5

4

MAZE CRAZE

Start
Here

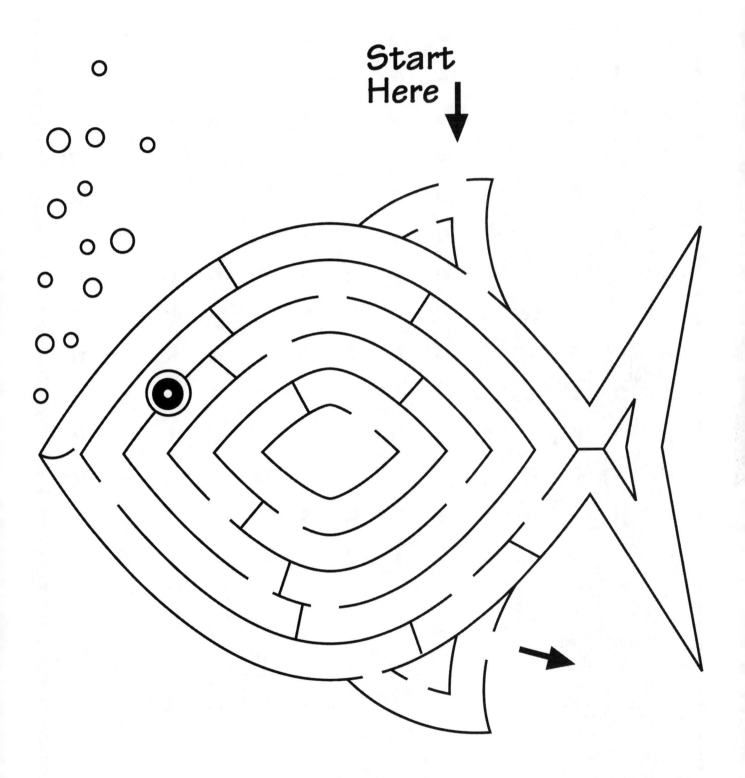

223

Word Search

```
y u y c g a a z k o j k u t
a u s t r a l i a u r n o a
z c g o y d o q t h y q a s
j k c c r o c o d i l e a m
j a c l x g h f z w l k c a
t n j i k q v f j o a u x n
u g r r o d j d n m h p v i
r a u w a c r b x b y k p a
n r e w l o e i e a c k p n
f o i z a v o t f t s j x d
g o k c o r a l r e e f f e
n s o g r o l i d o i k s v
h p l a t y p u s n z n p i
a b o r i g i n e a z y m l
```

aborigine
Australia
coral reef
crocodile
kangaroo

koala
platypus
tasmanian devil
wombat

224

FIND THE TWINS

Which two are exactly alike?

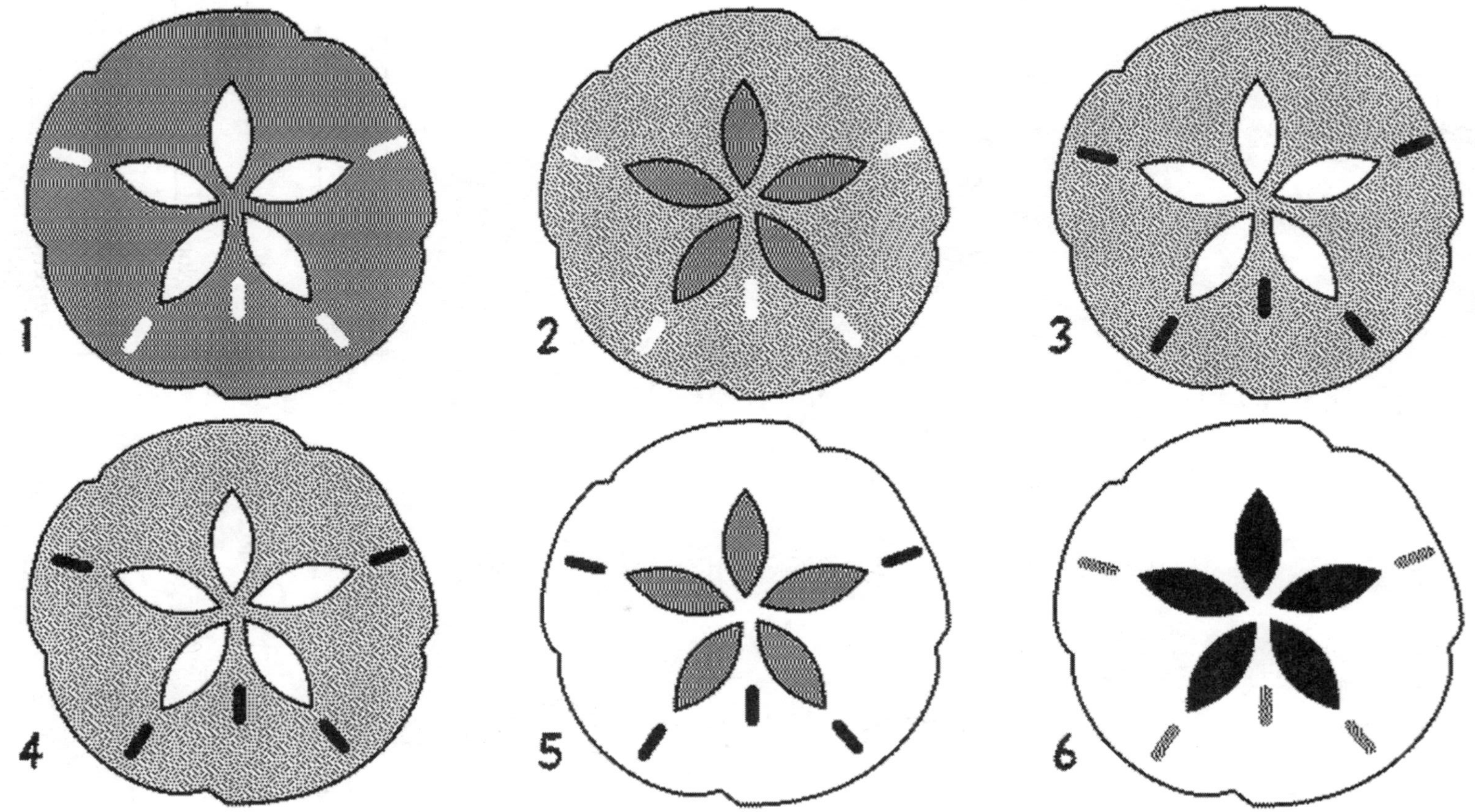

3

6

2

5

1

4

New Zealand

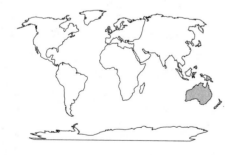

226

Population_____

Capital City_____

Religion_____

Type of Government_____

Currency_____

Language_____

What are the people called?_____

New Zealand

New Zealand is an island country 1,193 miles southeast of Australia. The two main islands are separated by Cook Strait, named after British explorer Captain James Cook.

The first settlers of New Zealand were Polynesian people known as Maoris. Today, Maoris make up less than 10% of New Zealand's population.

New Zealand's exports include wool, meat and dairy products, fruits, and vegetables.

People/History
- ❑ Margaret Mahy (author from New Zealand)
  - *My Mysterious World* by Margaret Mahy

General Reference Books
- ❑ *Take a Trip to New Zealand* by Geoff Burns
- ❑ *New Zealand* by Akinobu Yanagi
- ❑ Pages 74&75, *Children Just Like Me* by Barnabas & Anabel Kindersley

Literature
- ❑ Chapter 6 - *Missionary Stories with the Millers*
- ❑ *The Rattlebang Picnic* by Margaret Mahy
- ❑ *The Great White Man-Eating Shark* by Margaret Mahy
- ❑ *17 Kings and 42 Elephant* by Margaret Mahy
- ❑ *A Summery Saturday Morning* by Margaret Mahy

Science
- ❑ Fruits and vegetables
  - *Cool as a Cucumber, Hot as a Pepper* by Meredith Sayles Hughes
  - *Eat the Fruit, Plant the Seed* by Millicent E. Selsam
  - *Eating the Alphabet* by Lois Ehlert
  - *What's In the Names of Fruit* by Peter Limburg
  - *Growing Food* by Claire Llewellyn

  Notebook Suggestions:
  1. Name three crops that grow best in warm weather.
  2. Name three crops that grow best in cool weather.
  3. What is composting?
  4. Draw and label the parts of a plant.
  5. Draw and label the parts of a fruit.

Vocabulary
Maoris   haka   kumara

Music/Art/Projects

1. Color the flag of New Zealand.

    The flag of New Zealand is red, white, and blue. The British Union flag is in the upper left-hand corner signifying New Zealand's relationship with Great Britain. Four red stars outlined with white represent the Southern Cross constellation

2. Color or make a map of New Zealand.

3. Make Kumara and Apple Casserole.

    4 boiled sweet potatoes, peeled and cubed

    4 apples, peeled, cored, thinly sliced

    4 tablespoons butter

    1/2 cup brown sugar

    1/2 teaspoon salt

    Directions: Gently mix first four ingredients in a large bowl. Transfer to baking dish and dot with butter. Bake in a 350° oven for 1 hour. Serve as a side dish or dessert.

4. Make Rock Cakes.

    1 cup all-purpose flour

    1/2 cup each butter, sugar, raisins

    1 egg, lightly beaten

    1/2 teaspoon salt

    2 tablespoons orange marmalade

    Directions: Mix flour and salt in mixing bowl, blend in butter until like fine bread crumbs.
    Add sugar, raisins, and marmalade, mix well with spoon. Add egg until mixture is stiff and "rocky".
    Pull off golf ball-sized chunks of dough and place on cookie sheet 1-inch apart. Bake in 425∞ oven for 15 minutes. Remove from cookie sheet. Turn cakes upside down to cool. Serve with tea.

Internet Resources

❏ www.odci.gov/cia/publications/factbook/geos/nz.html

❏ www.infoplease.com/ipa/A0107834.html

❏ www.maori.org.nz/

❏ www.coolkids.co.nz/

❏ http://teacher.scholastic.com/glokid/zealand/

❏ http://library.christchurch.org.nz/

Bible

❏ Fruits and vegetables
   - Genesis 1:9-13
   - Genesis 3
   - Genesis 4:1-15
   - Psalm 1
   - Matthew 7:15-20
   - Mark 4:1-20
   - John 15:1-8
   - Galatians 5:22-23

❏ *The Rattlebang Picnic*
   - Psalm 127:3

❏ *The Great White Man-Eating Shark*
   - Proverbs 21:26
   - Isaiah 56:11

Flag of New Zealand

229

FIND THE TWINS

Which two are exactly alike?

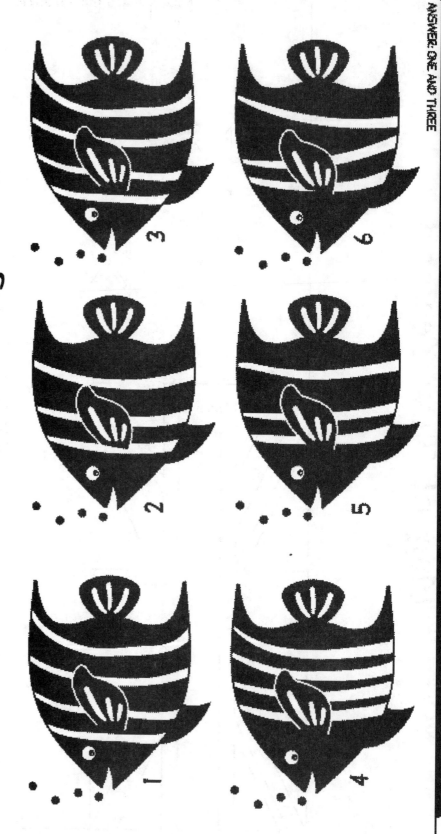

230

Appendix

Geography Dictionary

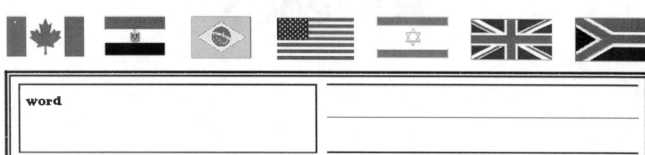

word

Draw a picture of what your word means.

Biography Report

What is my name? _____

When was I born? _____ When did I die? _____

Where did I live? _____

Who were my parents? _____

How many brothers and sisters did I have? _____

Who did I marry? _____

How many children did we have? _____

What kind of work did I do? _____

_____

Why am I important? _____

_____

_____

What mistakes did I make? _____

_____

_____

What did you learn from me? _____

_____

_____

Country Report

Country: _____

Continent: _____

Capital: _____

Language: _____   Put a picture here

What kind of houses do the people live in? _____

_____

What kind of food do the people eat? _____

_____

Geographical features: mountains   deserts   plains   coast

Climate: hot   cold   moderate   tropical

Major rivers and lakes: _____

Famous people: _____

_____

Animals: _____

_____

Would you like to visit this country?   Yes   No

Why? _____

_____

Animal Report

Name of the animal: _____

Where does the animal live? _____

What is its home like? _____

_____

What does it eat? _____

_____

This animal is awake during the: Day   Night

What kind of climate does this animal like? _____

_____

What special features did God give this animal? _____

_____

_____

Draw a picture of this animal.

Answer Key

Page: 10
Insects
1. head, thorax, abdomen
2. exoskeleton
3. to dig & to feed
4. Ants live and work together in large groups.
5. North Pole and South Pole
6. colony
8. by rubbing their wings together
9. with a special body part on their front legs

Page: 11
Butterflies
1. egg, chrysalis, pupa, butterfly
2. head, thorax, abdomen
3. to smell
4. scales
5. leaves and fruit
6. nectar

Page: 13
1. North America
2. South America
3. Europe
4. Asia
5. Africa
6. Australia

Page: 22
Ducks
1. webbed feet
2. from the water and areas surrounding it
3. layer of soft, fluffy feathers under the top feathers

Pandas
1. eating
2. to hold bamboo

Page: 23
Sloth Bear
1. two
2. termites and bees
3. nose

Orangutans
1. forest man
2. reddish- brown/orange
3. platforms in trees
4. as blankets

Snow Leopard
1. a purring sound
2. over hunting

Page: 37
Elephants
1. cow
2. trunk
3. leaves, grass, bark, berries
4. to spray water, pick up objects
5. An African elephant is bigger and has larger ears
6. Africa and India
7. Lions and hyenas will attack baby elephants, but adult elephants are safe from most predators.
8. human poachers

Tigers
1. cat family
2. Asia
3. deer, antelope, wild pigs, etc.
4. how to hunt and survive

Water Buffalo
1. water
2. to help protect themselves from insects
3. They pull a plow through water.

Page: 38
Chevrotain
1. long tusk like teeth
2. in the tropical forests of India & Southeast Asia
3. fruit, leaves, twigs, grass
4. mouse deer

Page: 45
Donkey
1. bigger ears, steadier, more patient
2. braying

Page: 54
Bears
1. smell, hearing
2. both plants and animals
3. winter sleep
4. by eating a lot to gain weight

Wolves
1. deer, elk, moose, etc.
2. strongest male
3. by lowering their ears and putting their tail between their legs
4. to cool down

Page: 59
Sheep
1. wool, milk, meat
2. to make clothing, blankets, rugs

Page: 63
Across
2. Sherwood
3. Thames
4. Big Ben
5. petrol
6. mate
8. Elizabeth

Down
1. London
2. Shakespeare
7. sheep

236

Page : 66

Answers will vary

Page : 71

Across
1. Paris
5. Eiffel Tower
6. Joan
7. Curie

Down
1. perfume
2. Monet
3. Pasteur
4. French

Page : 80

Across
1. Venus
3. Mars
4. Uranus
7. Milky Way
8. Jupiter
9. Saturn
10. Mercury
12. asteroid
13. planet
14. comet

Down
2. Neptune
3. moon
5. solar system
6. star
11. earth
13. Pluto

Page : 98

Cork
1. bark of a tree
2. answers vary
3. 300-400 years
4. 50 years

Page : 101
1. Spanish or Spaniard
2. Spanish
3. Portugese
4. Portugese

Page : 108

Deer and Reindeer
1. 300-600 lbs.
2. wooded areas
3. grass, leaves, plants
4. snow white tail
5. antlers
6. nuts, berries, tree buds, twigs
7. They have white spots.

Page : 113

Penguins
1. rookery
2. as flippers
3. 18 minutes
4. by keeping the egg or chick on the top of their feet

Page : 117

Wolverine
1. weasel
2. it never freezes
3. They will chase a bear or a mountain lion away from its food.
4. in early summer

Page : 118

Arctic Fox
1. brown
2. white
3. to protect itself from temperature and predators
4. small snout and ears
5. starts to shiver

Seals
1. flippers
2. oily fur, blubber
3. sea lion
4. white
5. to help hide them from polar bears

Orca (Killer Whale)
1. dolphin
2. black and white
3. over 100 lbs.
4. pods
5. by making sounds

Snowy Owls
1. feathers
2. their eyes cannot move
3. lemmings
4. on the ground

Page : 119

Polar Bears
1. thick fur, layer of fat
2. to help them swim
3. two years

Page : 126

Moose
1. 18-20
2. 35 mph
3. 45-58 mph
4. 11-16 lbs

Page : 127

Pigs (Chesters Barn)
1. by lying in the mud
2. snout
3. four
4. cheese, butter, milk,
5. scratch in the dirt to get food

Page :134

Across
2. horse
4. cat
6. duck
9. rooster
10. pig

Down
1. chicken
2. sheep
3. cow
4. turkey
5. dog

Page : 139
Buffalo
1. bison
2. provided food, hides & fuel
3. Buffalo can see 1/2mile, smell 1 mile, & hear 500 ft.
4. up to 35 mph

Rabbit
1. rodent
2. ears

Beaver
1. incisors
2. to steer in the water
3. a lodge
4. trees, logs, etc.
5. below

Otters
1. in water
2. webbed
3. barking, chirping or growling

Page : 140
Squirrels
1. cling to bark and balance
2. to have food in winter

Skunk
1. black & white fur, smell
2. cat
3. forests and grasslands
4. eat insects, rats and mice
5. raises tail and hisses

Raccoons
1. to develop fat for winter
2. a kit
3. they prefer food moist

Porcupine
1. bunches of hair grown close together
2. they are peaceful creatures
3. during day
4. raises quills on back

Page : 141
Cougars
1. mountain lion
2. Answers vary

Prairie dogs
1. rodent
2. sunflower

Manatees
1. sea cow
2. blue-gray
3. 10-13 feet

Page : 148
Across
6. Mississippi
8. Columbus
11. Hawaii
12. Lincoln
14. Georgia
15. Kentucky

Down
1. Arizona
2. Washington
3. Virginia
4. Missouri
5. Texas
7. Delaware
9. fifty
10. Florida
13. New jersey

Page : 157
Across
3. ocho
5. sies
8. cuatro
9. dies

Down
1. tres
2. uno
4. cinco
6. siete
7. nueve
9. Dos

Page : 162
Anteater
1. ants and termites
2. to open ant nests
3. it is sticky
4. in a hidden place

Chinchilla
1. thick, soft, blue-gray color
2. Andes Mountains
3. from plants

Page : 163
Giant armadillo
1. near rivers in eastern part of South America
2. "little armored one"
3. hard body plates
4. digging burrows and finding food

Jaguar
1. rainforests, mountains & woods
2. for their fur
3. yes

Sloth
1. upside down
2. 21-29 inches
3. because it moves slow

Birds
1. beak
2. breast
3. crown
4. flight feathers
5. flamingo, ostrich,
6. hummingbird
7. woodpecker
8. Answers vary
9. Answers vary

Page : 186
Giraffe
1. over 18 feet tall
2. tree leaves
3. they spread apart front legs

238

Page : 187

Apes and monkeys
1. monkeys have tails, apes don't
2. both stand on two legs both have human looking hands and feet.

Chimpanzees
1. ape
2. in rainforest
3. fruit, leaves, insects
4. make and use tools

Gorillas
1. adult male leader, females & their babies
2. family
3. Central Africa
4. on sleeping platforms on the ground or in trees
5. leaves and fruit
6. 4-5 lbs

Hippopotamus
1. by streams and marshes in Africa
2. grass, roots, reeds
3. calf
4. thick, bluish-gray
5. rest and sleep in water

Page : 188

Lions
1. to protect them when they fight
2. at night
3. pride
4. cub

Ostrich
1. 8 feet tall
2. 45 mph
3. on dry grassy plains and sandy deserts
4. leaves, seeds, flowers, insects and small animals
5. to move around eggs in nest

Rhino
1. for digging and defense
2. bushes and small trees
3. smell

Zebra
1. deserts and grasslands in eastern and southern Africa
2. run away
3. horse
4. eating

Page : 192

Across
1. ape
4. gorilla
6. monkey
9. chimp
10. hippo

Down
1. zebra
2. giraffe
5. lion
7. ostrich
8. rhino

Page : 218

Kangaroo
1. for balance
2. open forest and bush country of Australia
3. mob
4. joey
5. carry baby in pouch

Wombats
1. digging and burrowing
2. 75 pounds
3. at night
4. roots and leaves
5. up to 25 years

Koalas
1. eucalyptus tree
2. to go to another tree
3. with sharp claws
4. both carry young in the pouch
5. six months
6. 18 hours

Page : 219

Crocodile
1. to cool its body
2. by dragging it under water & rolling over and over until prey drowns

Tasmanian devil
1. from its bad temper and loud throaty growl
2. temper
3. remains of dead animals
4. 24 inches

Duck-billed platypus
1. answers will vary
2. digging burrows and getting food
3. with her tail
4. snails, worms, shrimp, small fish

Page : 227

Fruits and Vegetables
1. answers will vary
2. answers will vary
3. using a mixture of rotten vegetables, dirt, and grass for fertilizer

239

Resources

Considering God's Creation   2nd-7th grade

A great science program for teaching from a Christian World View. Instead of a dry textbook you and your students will love this thoughtfully planned out unit study. Everything is laid out for you in the lesson plans down to what you do and say. Each lesson has a scripture to support its theme, hands-on project, vocabulary, questions, and assignments for a student notebook. You can teach creation science in an entertaining way with "Evolution Stumpers" – arguments that refute evolution. Complete set includes 272 page Student Workbook, 112 page Teacher Manual, and CD with 23 songs. Extra student workbooks are available at a savings. Considering God's Creation is foundational for the Science assignments in *Galloping the Globe*. 29.95, extra student workbook, 13.95.

Uncle Josh's Outline Map Book or CD-ROM   All ages

Would you like more outline maps to enhance your geography and history studies? Uncle Josh, who designed all the maps in *Galloping the Globe*, has created a whole set of reproducible outline maps for home, school, and office. The quality digital map art includes rivers and surrounding boundaries. This is by far one of the best set of outline maps you'll find. Over 100 maps are available in your choice of reproducible book (112 pages, 19.95), or CD-ROM (26.95) usable by all computers that can run Acrobat Reader.

Geography Terms Chart   All ages

This color, laminated, composite landscape picture is designed to help students see and understand the earth's topographical physical features. Geographical terms are labeled right on the picture, plus a simple glossary of 154 terms on the back makes looking up geography vocabulary a cinch! Such vivid colors, you can use as a place mat! Great resource for the Geography Dictionary in *Galloping the Globe*. 11" x 15" laminated, 4.75.

Trail Guide to World Geography   2nd-10th grade

Students learn about their world continent by continent with 5-minute daily drills, mapping, building a geography notebook, and choosing from a wide variety of projects. This "week one, day one" kind of teacher's manual requires very little teacher preparation and is very flexible. Assign as much or as little as YOU decide. Daily drills are offered at 3 different levels for versatility and multi-year usage. All you need to get started is this book, an atlas, and outline maps. Uses *Geography Through Art* for added culture through hands-on art from throughout the world. 128 pages, 18.95.

Trail Guide to U.S. Geography   2nd-10th grade

In the same format as *Trail Guide to World Geography* described above, this book explores each of the 50 states with daily geography questions that can be answered using an atlas or almanac. Includes additional mapping, notebooking, and a wide selection of additional assignment choices for each state. 128 pages, 18.95.

Geography Through Art   All ages

A great way to learn about the world and the U.S. is through art. Complete instructions for art projects including, sculpture, drawing, pinata, and oh so many more. Organized by continent, all instructions for art projects in the *Trail Guide* series are in this book. 135 pages, 17.95.

Children's Atlases

Jr. Classroom Atlas is recommended for students in 2nd through 4th grade. 48 pages, 7.95.
Classroom Atlas is an excellent student atlas for 4th though 8th grade. 112 pages, 9.95.

To order any of these resources contact Geography Matters at 800-426-4650 or log onto the web site.

240   **www.geomatters.com**